Section Review Worksheets

HOLT
Physics

$$\sin \theta_c = \frac{n_r}{n_i}$$

PHYSICS INTERACTIVE TUTOR

SERWAY · FAUGHN

HOLT, RINEHART AND WINSTON

Harcourt Brace & Company

Austin · New York · Orlando · Atlanta · San Francisco · Boston · Dallas · Toronto · London

Holt Physics

Section Review Worksheets

This workbook consists of review and reinforcement activities that focus on key skills or concepts from a section of the *Holt Physics* text.

Graph Skills challenge students to make the connection between physics principles, equations, and their visual representation in a graph.

Diagram Skills bridge the gap between a real, physical situation, and the diagram that simplifies it so that key physics principles and equations can be applied.

Math Skills provide additional practice linking mathematical operations with chapter content.

Concept Reviews reinforce fundamental knowledge from a section of the text.

Mixed Review include items that check student's comprehension of a variety of concepts from throughout the chapter.

Worksheet Authors

Phillip G. Bunce
James Bowie High School
Austin, TX

Judith R. Edgington, Ph. D.
Physics/Science Education Consultant and Curriculum Designer
Austin, TX

Printed in the United States of America

ISBN 0-03-051869-5

8 9 10 11 018 09 08 07 06

Contents

Section
1-1

HOLT PHYSICS
Concept Review

What is Physics?

1. Which areas of physics deal with the following?

    a. how fast things move _____

    b. how the shape of a cave affects an echo _____

    c. which sunglasses are best for cutting the glare on a ski slope _____

    d. how the cooling system in a refrigerator works _____

    e. what lightning is _____

    f. how energy is produced by the sun _____

2. Laws governing speed limits on highways are determined by a majority vote by citizens of a state or their representatives. Compare this democratic procedure to the way scientific laws are established with regard to the following questions. Explain your reasoning.

    a. Can scientific laws be changed by a vote?

    _____

    _____

    _____

    b. Can the speed of light be legislated?

    _____

    _____

    _____

    c. Can scientists from other countries change what physicists in the United States think?

    _____

    _____

    _____

Section

1-2

HOLT PHYSICS

Math Skills

Measurements in Experiments

Power	Prefix	Abbreviation
10^{-18}	atto-	a
10^{-15}	femto-	f
10^{-12}	pico-	p
10^{-9}	nano-	n
10^{-6}	micro-	μ
10^{-3}	milli-	m
10^{-2}	centi-	c

Power	Prefix	Abbreviation
10^{-1}	deci-	d
10^{1}	deka-	da
10^{3}	kilo-	k
10^{6}	mega-	M
10^{9}	giga-	G
10^{12}	tera-	T
10^{15}	peta-	P
10^{18}	exa-	E

1. How many picoseconds are there in 1 Ms? _____

2. How many micrograms make 1 kg? _____

3. How many nanometers are there in 1 cm? _____

4. Rewrite the following quantities in scientific notation without prefixes.

    a. 3582 gigabytes _____

    b. 0.0009231 milliwatts _____

    c. 53657 nanoseconds _____

    d. 5.32 milligrams _____

    e. 88900 megahertz _____

    f. 0.00000083 centimeters _____

5. Rewrite the following quantities in units with SI prefixes.

    a. 36582472 g _____

    b. 0.000000452 m _____

    c. 53236 V _____

    d. 4.62×10^{-3} s _____

6. Express the measurement 4.29478416 kg with 8, 6, 4, and 2 significant figures.

_____ _____

_____ _____

HOLT PHYSICS
Math Skills

The Language of Physics

1. Calculate the following products and quotients without using a calculator.

   a. $(3.0 \times 10^5) \times (2.0 \times 10^3)$ _____

   b. $(3.0 \times 10^5) \div (2.0 \times 10^3)$ _____

   c. $(3.0 \times 10^2) \div (2.0 \times 10^5)$ _____

   d. $(3.0 \times 10^{-2}) \times (2.0 \times 10^5)$ _____

   e. $(3.0 \times 10^{-2}) \div (2.0 \times 10^{-5})$ _____

   f. $(3.0 \times 10^{-2}) \times (2.0 \times 10^{-5})$ _____

2. Round off the following numbers to one figure, then find their order of magnitude.

   a. 3.7×10^5 _____

   b. 6.1×10^5 _____

   c. 8.2×10^{-9} _____

   d. 0.000067 _____

   e. 7439262 _____

   f. 0.0006739 _____

3. Find the order of magnitude of the following results without using a calculator.

   a. 97×192 _____

   b. $96.8639 \div 883.3525$ _____

4. **a.** Estimate the width and height in centimeters of a half-gallon milk container. Show your assumptions and your work.

   b. Use your numbers to obtain a rough estimate of the volume of milk in a half-gallon container. _____

   c. The volume of a half-gallon is about 1890 cm³. How close was your estimate? _____

Chapter

HOLT PHYSICS

Mixed Review

1

The Science of Physics

Power	Prefix	Abbreviation
10^{-18}	atto-	a
10^{-15}	femto-	f
10^{-12}	pico-	p
10^{-9}	nano-	n
10^{-6}	micro-	μ
10^{-3}	milli-	m
10^{-2}	centi-	c

Power	Prefix	Abbreviation
10^{-1}	deci-	d
10^{1}	deka-	da
10^{3}	kilo-	k
10^{6}	mega-	M
10^{9}	giga-	G
10^{12}	tera-	T
10^{15}	peta-	P
10^{18}	exa-	E

1. Convert the following measurements to the units specified.

    a. 2.5 days to seconds _____

    b. 35 km to millimeters _____

    c. 43 cm to kilometers _____

    d. 22 mg to kilograms _____

    e. 671 kg to micrograms _____

    f. 8.76×10^{7} mW to gigawatts _____

    g. 1.753×10^{-13} s to picoseconds _____

2. According to the rules given in Chapter 1 of your textbook, how many significant figures are there in the following measurements?

    a. 0.0845 kg _____

    b. 37.00 h _____

    c. 8 630 000.000 mi _____

    d. 0.000 000 0217 g _____

    e. 750 in. _____

    f. 0.5003 s _____

Chapter

1 HOLT PHYSICS
Mixed Review *continued*

3. **Without calculating the result,** find the number of significant figures in the following products and quotients.

   a. 0.005032×4.0009 _____

   b. $0.0080750 \div 10.037$ _____

   c. $(3.52 \times 10^{-11}) \times (7.823 \times 10^{11})$ _____

4. Calculate $a + b$, $a - b$, $a \times b$, and $a \div b$ with the correct number of significant figures using the following numbers.

   a. $a = 0.005\ 078$; $b = 1.0003$

   $a + b =$ _____ $a - b =$ _____

   $a \times b =$ _____ $a \div b =$ _____

   b. $a = 4.231\ 19 \times 10^{7}$; $b = 3.654 \times 10^{6}$

   $a + b =$ _____ $a - b =$ _____

   $a \times b =$ _____ $a \div b =$ _____

5. Calculate the area of a carpet 6.35 m long and 2.50 m wide. Express your answer with the correct number of significant figures.

   _____

6. The table below contains measurements of the temperature and volume of an air balloon as it heats up.

   In the grid at right, sketch a graph that best describes these data.

Temperature (°C)	Volume (m³)
2	0.0502
27	0.0553
52	0.0598
77	0.0646
102	0.0704
127	0.0748
152	0.0796

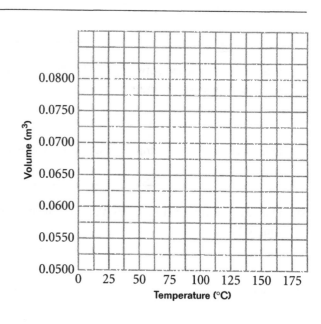

Section

2-1

HOLT PHYSICS

Graph Skills

Displacement and Velocity

A minivan travels along a straight road. It initially starts moving toward the east. Below is the position-time graph of the minivan. Use the information in the graph to answer the questions.

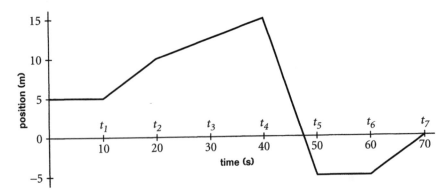

1. Does the minivan move to the east? If so, during which time interval(s)?

   _____

2. Does the minivan move to the west? If so, during which time interval(s)?

   _____

3. Is the minivan's speed between t_1 and t_2 greater than, less than, or equal to its speed between t_2 and t_3?

   _____

4. Is the minivan's speed between t_4 and t_5 greater than, less than, or equal to its speed between t_6 and t_7?

   _____

5. Does the minivan ever stop completely? If so, at which time(s)?

   _____

6. Does the minivan ever move with a constant velocity? If so, at which time(s)?

   _____

7. What is the total displacement of the minivan during the trip?

   _____

HOLT PHYSICS
Math Skills

Acceleration

A car is traveling down a straight road. The driver then applies the brake, and the car decelerates with a constant acceleration until it stops. Refer to the equations below to answer the questions.

$$\Delta x = \frac{1}{2}(v_i + v_f)\Delta t \qquad v_f = v_i + a(\Delta t)$$
$$\Delta x = v_i(\Delta t) + \frac{1}{2}a(\Delta t)^2 \qquad v_f^2 = v_i^2 + 2a\Delta x$$

1. What is the car's final speed v_f? Explain your answer.

_____

_____

_____

2. You are given the distance the car travels and the length of time it takes for the car to come to a complete stop after the driver applies the brakes. What is the expression for the car's initial speed?

_____

3. You are given the car's initial speed and the length of time it takes for the car to come to a full stop after the driver applies the brakes. What is the expression for the magnitude of the car's acceleration?

_____

4. You are given the car's initial speed and the distance the car travels before it comes to a complete stop after the driver applies the brakes. What is the expression for the magnitude of the car's acceleration?

_____

5. You are given the magnitude of the car's acceleration and the length of time it takes for the car to come to a full stop after the driver applies the brakes. What is the expression for the initial speed of the car, and what is the expression for the distance it traveled before it came to a complete stop?

_____

Section
2-3

HOLT PHYSICS
Math Skills

Falling Objects

A juggler throws a ball straight up into the air. The ball remains in the air for a time Δt before it lands back in the juggler's hand.

1. Answer the following questions in terms of Δt and g.

   a. What is the acceleration of the ball during the entire time the ball is in the air?

   _____

   b. With what speed did the juggler throw the ball into the air? (Hint: What is the total displacement of the ball during the time it is in the air?)

   _____

   c. How much time elapsed before the ball reached its maximum height?

   _____

   d. How high above the point of release did the ball rise?

   _____

   e. Suppose that the ball remains in the air for an amount of time equal to $2\Delta t$. How much higher would the ball rise in the air?

   _____

2. Assume that the ball was in the air for 2.4 s. Answer the following questions:

   a. What is the acceleration of the ball during the entire time the ball is in the air?

   _____

   b. With what speed did the juggler throw the ball into the air?

   _____

   c. How much time elapsed before the ball reached its maximum height?

   _____

   d. How high above the point of release did the ball rise?

   _____

Chapter

2

HOLT PHYSICS

Mixed Review

Motion in One Dimension

1. During a relay race along a straight road, the first runner on a three-person team runs d_1 with a constant velocity v_1. The runner then hands off the baton to the second runner, who runs d_2 with a constant velocity v_2. The baton is then passed to the third runner, who completes the race by traveling d_3 with a constant velocity v_3.

   a. In terms of d and v, find the time it takes for each runner to complete a segment of the race.

   Runner 1 _____   Runner 2 _____   Runner 3 _____

   b. What is the total distance of the race course?

   _____

   c. What is the total time it takes the team to complete the race?

   _____

2. The equations below include the equations for straight-line motion. For each of the following problems, indicate which equation or equations you would use to solve the problem, but do not actually perform the calculations.

$$\Delta x = \tfrac{1}{2}(v_i + v_f)\Delta t \qquad \Delta x = \tfrac{1}{2}(v_f)\Delta t$$
$$\Delta x = v_i(\Delta t) + \tfrac{1}{2}a(\Delta t)^2 \qquad \Delta x = \tfrac{1}{2}a(\Delta t)^2$$
$$v_f = v_i + a(\Delta t) \qquad v_f = a(\Delta t)$$
$$v_f^2 = v_i^2 + 2a\Delta x \qquad v_f^2 = 2a\Delta x$$

   a. During takeoff, a plane accelerates at 4 m/s^2 and takes 40 s to reach takeoff speed. What is the velocity of the plane at takeoff?

   _____

   b. A car with an initial speed of 31.4 km/h accelerates at a uniform rate of 1.2 m/s^2 for 1.3 s. What is the final speed and displacement of the car during this time?

   _____

3. Below is the velocity-time graph of an object moving along a straight
path. Use the information in the graph to fill in the table below.

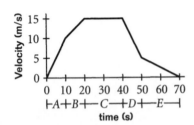

For each of the lettered intervals below, indicate the motion of the object
(whether it is speeding up, slowing down, or at rest), the direction of the
velocity (+, –, or 0), and the direction of the acceleration (+, –, or 0).

Time interval	Motion	v	a
A			
B			
C			
D			
E			

4. A ball is thrown upward with an initial velocity of 9.8 m/s from the top
of a building.

a. Fill in the table below showing the ball's position, velocity, and acceleration at the end of each of the first 4 s of motion.

Time (s)	Position (m)	Velocity (m/s)	Acceleration (m/s^2)
1			
2			
3			
4			

b. In which second does the ball reach the top of its flight?

_____

c. In which second does the ball reach the level of the roof, on the
way down?

_____

Section

3-1

HOLT PHYSICS
Diagram Skills

Introduction to Vectors

Use the following vectors to answer the questions.

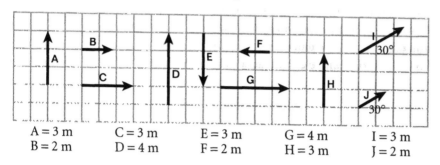

| A = 3 m | C = 3 m | E = 3 m | G = 4 m | I = 3 m |
| B = 2 m | D = 4 m | F = 2 m | H = 3 m | J = 2 m |

1. Which vectors have the same magnitude?

_____

2. Which vectors have the same direction?

_____

3. Which arrows, if any, represent the same vector?

_____

4. In the space provided, construct and label a diagram that shows the vector sum 2**A** + **B.** Construct and label a second diagram that shows **B** + 2**A.**

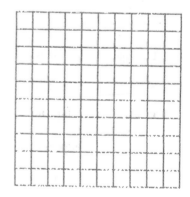

5. In the space provided, construct and label a diagram that shows the vector difference **A** – (**B**/2). Construct and label a second diagram that shows (**B**/2) – **A.**

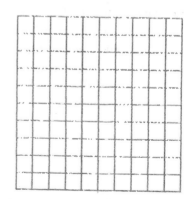

Section
3-2
HOLT PHYSICS
Diagram Skills
Vector Operations

One of the holes on a golf course lies due east of the tee. A novice golfer flubs his tee shot so that the ball lands only 64 m directly northeast of the tee. He then slices the ball 30° south of east so that the ball lands in a sand trap 127 m away. Frustrated, the golfer then blasts the ball out of the sand trap, and the ball lands at a point 73 m away at an angle 27° north of east. At this point, the ball is on the putting green and 14.89 m due north of the hole. To his amazement, the golfer then sinks the ball with a single shot.

1. In the space provided, choose a scale, then draw a sketch of the displacement for each shot the golfer made. Label the magnitude of each vector and the angle of each vector relative to the horizontal axis.

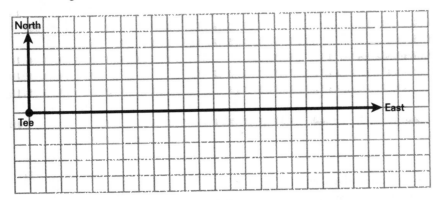

2. Use algebraic formulas to find the x and y components of each displacement vector.

Shot 1 x component _____ y component _____

Shot 2 x component _____ y component _____

Shot 3 x component _____ y component _____

Shot 4 x component _____ y component _____

3. Find the total distance (to the nearest meter) the golf ball traveled from the tee to the hole. Assume the golf course is flat. (Hint: Which component of each displacement vector contributes to the total displacement of the ball between the tee and the hole?)

_____

Section
3-3
HOLT PHYSICS
Math Skills

Projectile Motion

After a snowstorm, a boy and a girl decide to have a snowball fight. The girl uses a large slingshot to shoot snowballs at the boy. Assume that the girl fires each snowball at an angle θ from the ground and that the snowballs travel with an initial velocity of v_0.

1. In terms of the initial velocity, v_0, and the launch angle, θ, for what amount of time, Δt, will a snowball travel before it reaches its maximum height above the ground? (Hint: Recall that $v_f = 0$ when an object reaches its maximum height.)

_____

2. What is the maximum height, h, above the ground that a snowball reaches after it has been launched?

_____

3. What is the horizontal distance, x, the snowball has traveled when it reaches its maximum height?

_____

4. The range, R, is the horizontal distance traveled in *twice* the time it takes for an object to reach its maximum height. Using your answers from items 1 and 3, write an expression for the range in terms of v_0, θ, and g.

_____

5. If the initial velocity, v_0, equals 50.00 m/s, find the maximum height and range for each of the launch angles listed in the table below.

Launch angle	Maximum height (m)	Range (m)
15°		
30°		
45°		
60°		
75°		

Section
3-4

HOLT PHYSICS
Diagram Skills

Relative Motion

The water current in a river moves relative to the land with a velocity v_{WL}, and a boat is traveling on the river relative to the current with a velocity v_{BW}.

1. How is the velocity of the boat relative to the land (v_{BL}) related to v_{WL} and v_{BW}?

_____

2. Suppose that both the boat and the water current move in the same direction and that the boat is moving twice as fast as the current. Draw a vector diagram to determine the velocity of the boat relative to the land, v_{BL}.

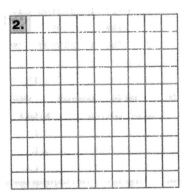

3. Suppose that the boat travels in the opposite direction of the current and that the boat is moving twice as fast as the current. Draw a vector diagram to determine the velocity of the boat relative to the land, v_{BL}.

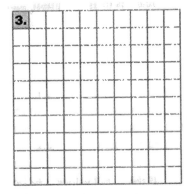

4. Suppose that the boat travels in a direction perpendicular to the current and that the boat is moving twice as fast as the current. Draw a vector diagram to determine the velocity of the boat relative to the land, v_{BL}.

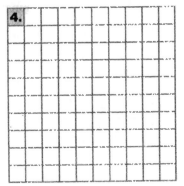

5. Assume that the boat travels with a speed of 4.0 km/h relative to the current and that the current moves due east at a speed of 2.0 km/h relative to the land. Determine the velocity of the boat relative to the land for each of the situations described in items 2–4.

   a. v_{BL} for item 2 _____

   b. v_{BL} for item 3 _____

   c. v_{BL} for item 4 _____

Chapter

3

HOLT PHYSICS

Mixed Review

Two-Dimensional Motion and Vectors

1. The diagram below indicates three positions to which a woman travels. She starts at position *A*, travels 3.0 km to the west to point *B*, then 6.0 km to the north to point *C*. She then backtracks, and travels 2.0 km to the south to point *D*.

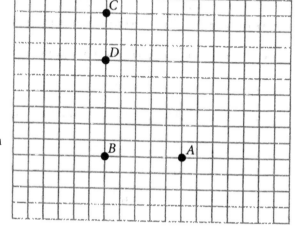

a. In the space provided, diagram the displacement vectors for each segment of the woman's trip.

b. What is the total displacement of the woman from her initial position, *A*, to her final position, *D*?

_____

c. What is the total distance traveled by the woman from her initial position, *A*, to her final position, *D*?

_____

2. Two projectiles are launched from the ground, and both reach the same vertical height. However, projectile B travels twice the horizontal distance as projectile A before hitting the ground.

a. How large is the vertical component of the initial velocity of projectile B compared with the vertical component of the initial velocity of projectile A?

_____

b. How large is the horizontal component of the initial velocity of projectile B compared with the horizontal component of the initial velocity of projectile A?

_____

c. Suppose projectile A is launched at an angle of 45° to the horizontal. What is the ratio, v_B/v_A, of the speed of projectile B, v_B, compared with the speed of projectile A, v_A?

_____

3. A passenger at an airport steps onto a moving sidewalk that is 100.0 m long and is moving at a speed of 1.5 m/s. The passenger then starts walking at a speed of 1.0 m/s in the same direction as the sidewalk is moving. What is the passenger's velocity relative to the following observers?

a. A person standing stationary alongside to the moving sidewalk.

_____

b. A person standing stationary *on* the moving sidewalk.

_____

c. A person walking alongside the sidewalk with a speed of 2.0 m/s and in a direction opposite the motion of the sidewalk.

_____

d. A person riding in a cart alongside the sidewalk with a speed of 5.0 m/s and in the same direction in which the sidewalk is moving.

_____

e. A person riding in a cart with a speed of 4.0 m/s and in a direction perpendicular to the direction in which the sidewalk is moving.

_____

4. Use the information given in item 3 to answer the following questions:

a. How long does it take for the passenger walking on the sidewalk to get from one end of the sidewalk to the other end?

_____

b. How much time does the passenger save by taking the moving sidewalk instead of walking alongside it?

_____

Section
4-1
HOLT PHYSICS
Diagram Skills

Changes in Motion

A large, square box of exercise equipment sits on a storeroom floor. A rope is tied around the box. Assume that if the box moves along the floor, there is a backward force that resists its motion.

1. Suppose that the box remains at rest. In the space provided, draw a free-body diagram for the box. Label each force involved in the diagram.

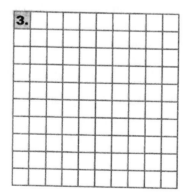

2. Suppose a warehouse worker moves the box by pulling the rope to the right horizontal to the ground. In the space provided, draw a free-body diagram for the box. Label each force involved in the diagram.

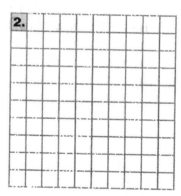

3. Suppose the warehouse worker moves the box by pulling the rope to the right at a 50° angle to the ground. In the space provided, draw a free-body diagram for the box. Label each force involved in the diagram.

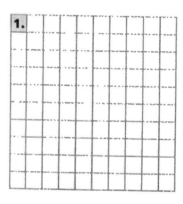

HOLT PHYSICS

4-2 Diagram Skills

Newton's First Law

A lantern of mass *m* is suspended by a string that is tied to two other strings, as shown in the figure below. The free-body diagram shows the forces exerted by the three strings on the knot.

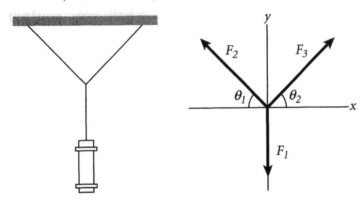

1. In terms of F_1, F_2, and F_3, what is the net force acting on the knot? (Hint: The lantern is in equilibrium.)

_____

2. Find the *x* and *y* components for each force acting on the knot. (Assume the positive directions are to the right and up.)

   String 1 (F_1)   *x* component _____   *y* component _____

   String 2 (F_2)   *x* component _____   *y* component _____

   String 3 (F_3)   *x* component _____   *y* component _____

3. In terms of F_1, F_2, and F_3, what is the net force acting on the knot in the *x* direction? in the *y* direction?

   $F_{x\ net}$ = _____

   $F_{y\ net}$ = _____

4. Assume that $\theta_1 = 30°$, $\theta_2 = 60°$, and the mass of the lantern is 2.1 kg. Find F_1, F_2, and F_3.

   F_1 = _____

   F_2 = _____

   F_3 = _____

Section

4-3

HOLT PHYSICS

Diagram Skills

Newton's Second and Third Laws

The figure on the left below illustrates a sled with a mass of *M* pulled horizontally along the ground by a force with a magnitude of *F*. A box with a mass of *m* lies on the sled and remains at rest relative to the sled. Assume there is friction between the surface of the sled and the box and between the surface of the ground and the sled. The figure on the right below shows the *force* diagram for this situation.

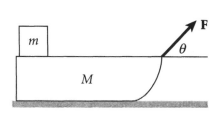 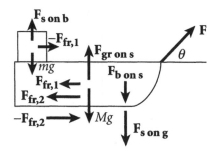

1. Identify any action-reaction pairs in the force diagram.

_____

2. Which of the forces shown would be included in the free-body diagram of the box?

_____

3. Which of the forces shown would be included in the free-body diagram of the sled?

_____

4. What is the net force on the box in the horizontal direction? _____

5. What is the net force on the box in the vertical direction? _____

6. What is the net force on the sled in the horizontal direction? _____

7. What is the net force on the sled in the vertical direction? _____

HOLT PHYSICS

4-4 Concept Review

Everyday Forces

A wooden box with a mass of 10.0 kg rests on a ramp that is inclined at an angle of 25° to the horizontal. A rope attached to the box runs parallel to the ramp and then passes over a frictionless pulley. A bucket with a mass of *m* hangs from the end of the rope. The coefficient of static friction between the ramp and the box is 0.50. The coefficient of kinetic friction between the ramp and the box is 0.35.

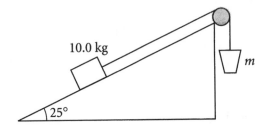

10.0 kg

m

25°

1. Suppose the box remains at rest relative to the ramp. What is the maximum magnitude of the friction force exerted on the box by the ramp?

_____

2. Suppose the box slides along the ramp. What is the maximum magnitude of the friction force exerted on the box by the ramp?

_____

3. Suppose the bucket has a mass of 2.0 kg.

   a. What is the friction force exerted on the box by the ramp?

   _____

   b. Does the box remain at rest relative to the ramp?

   _____

4. Suppose water is added to the bucket so that the total mass of the bucket and its contents is 6.0 kg.

   a. What is the friction force exerted on the box by the ramp?

   _____

   b. Does the box remain at rest relative to the ramp?

   _____

Chapter 4

HOLT PHYSICS
Mixed Review

Forces and the Laws of Motion

1. A crate rests on the smooth, horizontal bed of a pickup truck. For each of the situations described below, indicate (a) the motion of the crate relative to the ground, (b) the motion of the crate relative to the truck, and (c) whether the crate will hit the front wall of the truck bed, the back wall, or neither.

   a. Starting at rest, the truck *accelerates* to the right.

   _____

   b. The crate is at rest relative to the truck while the truck moves to the right with a constant velocity.

   _____

   c. The truck in item b slows down.

   _____

2. A ball with a mass of *m* is thrown through the air, as shown in the figure.

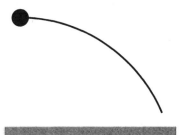

   a. What is the gravitational force exerted on the ball by Earth?

   _____

   b. What is the force exerted by Earth on the ball?

   _____

   c. If the surrounding air exerts a force on the ball that resists its motion, is the *total* force on the ball the same as the force calculated in part a?

   _____

   d. If the surrounding air exerts a force on the ball that resists its motion, is the *gravitational* force on the ball the same as the force calculated in part a?

   _____

3. Two blocks of masses m_1 and m_2, respectively, are placed in contact with each other on a smooth, horizontal surface. A constant horizontal force F to the right is applied to m_1. Answer the following questions in terms of F, m_1, and m_2.

   a. What is the acceleration of the two blocks?

   _____

   b. What are the horizontal forces acting on m_2?

   _____

   c. What are the horizontal forces acting on m_1?

   _____

   d. What is the magnitude of the contact force between the two blocks?

   _____

4. Assume you have the same situation as described in item 3, only this time there is a frictional force, F_k, between the blocks and the surface. Answer the following questions in terms of F, F_k, m_1, and m_2.

   a. What is the acceleration of the two blocks?

   _____

   b. What are the horizontal forces acting on m_2?

   _____

   c. What are the horizontal forces acting on m_1?

   _____

   d. What is the magnitude of the contact force between the two blocks?

   _____

Section

5-1

HOLT PHYSICS

Math Skills

Work

A crate with a mass of *m* is on a ramp that is inclined at an angle of 30° from the horizontal. A force with a magnitude of *F* directed parallel to the ramp is used to pull the crate with a constant speed up the ramp a distance of *d*.

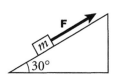 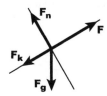

1. What is the work done on the crate by the applied force *F*?

_____

2. What is the work done on the crate by the gravitational force exerted on the crate by Earth?

_____

3. What is the work done on the crate by the normal force, with a magnitude of F_n, exerted on the crate by the ramp? (Hint: recall that the normal force is perpendicular to the surface of the ramp.)

_____

4. What is the work done on the crate by the frictional force F_k?

_____

5. What is the total force acting on the crate?

_____

6. What is the work done on the crate by the total force?

_____

HOLT PHYSICS

Diagram Skills

Energy

As shown in the diagram, a block with a mass of m slides on a frictionless, horizontal surface with a constant velocity of v_i. It then collides with a spring that has a spring constant of k. The block fully compresses the spring, comes to rest briefly, and then moves in the opposite direction with a velocity of $-v_i$.

(a)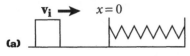

(b)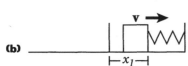

(c)

(d)

1. Examine the situation shown in part **(a)** of the diagram.

   a. What is the kinetic energy of the block? _____

   b. What is the potential energy associated with the block's position? _____

   c. What is the mechanical energy for this system? _____

2. Examine the situation shown in part **(b)** of the diagram.

   a. What is the kinetic energy of the block? _____

   b. What is the potential energy associated with the block's position? _____

   c. What is the mechanical energy for this system? _____

3. Examine the situation shown in part **(c)** of the diagram.

   a. What is the kinetic energy of the block? _____

   b. What is the potential energy associated with the block's position? _____

   c. What is the mechanical energy for this system? _____

4. Examine the situation shown in part **(d)** of the diagram.

   a. What is the kinetic energy of the block? _____

   b. What is the potential energy associated with the block's position? _____

   c. What is the mechanical energy for this system? _____

Section

5-3

HOLT PHYSICS

Diagram Skills

Conservation of Energy

A roller-coaster car with a mass of *m* moves along a smooth track as dia-
grammed in the graph below. The car leaves point *A* with no initial velocity
and travels to other points along the track. The zero energy level is taken as
the energy of point *A*.

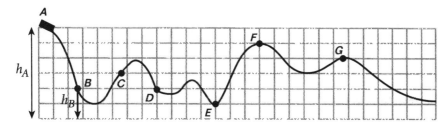

1. **a.** What is the car's kinetic energy at point *A*? _____

   b. What is the potential energy associated with the car at point *A*? _____

   c. What is the car's kinetic energy at point *B*? _____

   d. What is the potential energy associated with the car at point *B*? _____

2. **a.** What is the speed of the car at point *A*? _____

   b. What is the speed of the car at point *B*? _____

3. Assume the mass of the car is 65.0 kg and it starts at 30.0 m above the
ground. Use the graph above to find the kinetic energy, potential energy,
and velocity for points *C*, *D*, *E*, *F*, and *G* to complete the table.

Location	KE_A	PE_A	$KE_{location}$	$PE_{location}$	$v_{location}$
C					
D					
E					
F					
G					

4. For each location, what do you notice about the sum $KE_A + PE_A$ compared with the sum $KE_{location} + PE_{location}$?

_____

_____

_____

Section
5-4

HOLT PHYSICS
Concept Review

Work, Energy, and Power

A man accidentally knocks a flowerpot off a high window ledge. The flowerpot drops straight down under the influence of gravity.

1. What is the velocity of the flowerpot as it falls?

_____

2. What is the distance the flowerpot falls?

_____

3. What is the force acting on the flowerpot as it falls?

_____

4. What is the work down on the flowerpot as it falls?

_____

5. Assume the flowerpot has a mass of 5.0 kg and drops a total distance of 15.0 m. In the space provided, graph the work done on the flowerpot as a function of time.

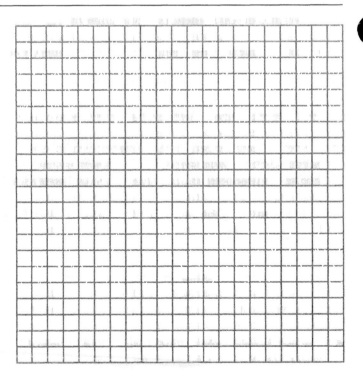

HOLT PHYSICS
5 Mixed Review

Work and Energy

1. A ball has a mass of 3 kg. What is the work done on this ball by the gravitational force exerted by Earth if the ball moves 2 m along each of the following directions?

   a. downward (along the force) _____

   b. upward (opposite the force) _____

2. A stone with a mass of m is thrown off a building. As the stone passes point A, it has a speed of v_A at an angle of θ to the horizontal. The stone then travels a vertical distance h to point B, where it has a speed v_B.

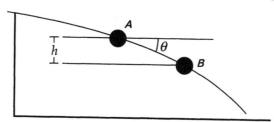

   a. What is the work done on the stone by the gravitational force due to Earth while the stone moves from A to B?

   _____

   b. What is the change in the kinetic energy of the stone as it moves from A to B?

   _____

   c. What is the speed v_B of the stone in terms of v_A, g, and h?

   _____

   d. Does the change in the stone's speed between A and B depend on the mass of the stone?

   _____

   _____

   e. Does the change in the stone's speed between A and B depend on the angle θ?

   _____

3. An empty coffee mug with a mass of 0.40 kg gets knocked off a tabletop 0.75 m above the floor onto the seat of a chair 0.45 m above the floor. Assume that the gravitational potential energy, PE_g, is measured using the floor as the zero energy level.

a. What is the initial gravitational potential energy associated with the mug's position on the table?

_____

b. What is the final gravitational potential energy associated with the mug's position on the chair seat?

_____

c. What was the work done by the gravitational force as it fell from the table to the chair?

_____

d. Suppose that zero level for the energy was taken to be the ceiling of the room rather than the floor. Would the answers to items a to c be the same or different?

_____

_____

4. A carton of shoes with a mass of m slides with an initial speed of v_i m/s down a ramp inclined at an angle of 23° to the horizontal. The carton's initial height is h_i, and its final height is h_f, and it travels a distance of d down the ramp. There is a frictional force, F_k, between the ramp and the carton.

a. What is the initial mechanical energy, ME_i, of the carton? (Hint: Apply the law of conservation of energy.)

_____

b. If μ is the coefficient of friction between the ramp and the carton, what is F_k?

_____

c. Express the final speed, v_f, of the carton in terms of v_i, g, d, and μ.

_____

HRW material copyrighted under notice appearing earlier in this book.

1. A soccer ball with a mass of 0.950 kg is traveling east at 10.0 m/s. Using a ruler and a scale of 1.0 square per 1.0 kg•m/s, draw a vector representing the momentum of the soccer ball.

2. A force of 2.00×10^2 N directed south is exerted on the ball for 0.025 s. Using the technique you used in item 1, draw a vector representing the impulse on the soccer ball.

3. The final momentum of the soccer ball is the initial momentum plus the change in momentum. Add your vectors from the previous questions to draw the final momentum vector of the ball.

4. Use your scale (1.0 square = 1.0 kg•m/s) to find the magnitude of the final momentum.

_____

5. Using your value for final momentum and the mass given in item 1, find the final speed of the ball.

_____

6. How can you determine the angle at which the ball is traveling?

_____

_____

_____

7. Use the techniques you used in items 1–5 to find the final speed of a 0.150 kg baseball that initially travels east at 40.0 m/s and is then hit with a westward force of 1250 N over a 0.010 s interval.

_____

_____

_____

_____

Use this grid for items 1–6.

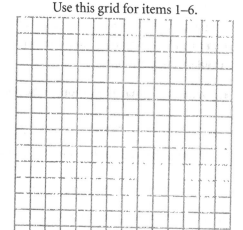

Use this grid for item 7.

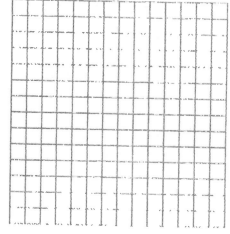

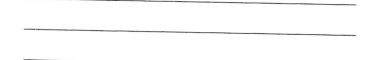

Section
6-2
HOLT PHYSICS
Concept Review

Conservation of Momentum

A radioactive nucleus is initially at rest. When it decays, it splits into two moving parts, one of which has exactly 50 times the mass of the other. Assume there are no external forces acting on the nucleus, and answer the following questions.

1. What is the total momentum of the nucleus before the fission (split) occurs?

   _____

2. What is the total momentum of the pieces after the event?

   _____

3. Assume the less massive particle moves east (0°). In words, compare the size and direction of the two momentum vectors.

   _____

   _____

   _____

4. Because the masses are different, the velocities must be different. Determine the ratio of the velocity of the small particle to the velocity of the large particle.

   _____

5. What generalization can you make about the relative velocities and the masses in this type of situation?

   _____

   _____

   _____

HOLT PHYSICS
Diagram Skills

Elastic and Inelastic Collisions

Use the following vectors to answer items 1–5.

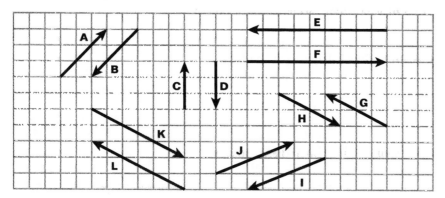

Consider a collision between two objects. Assume that the initial momentum of object 1 is represented by vector **A** ($p_{1,i} = A$) and the initial momentum of object 2 is represented by vector **B** ($p_{2,i} = K$).

1. In the space below, construct a vector diagram showing the total initial momentum just before the collision.

2. Which vector above represents the total initial momentum?

   _____

3. Which vector above represents the total final momentum?

   _____

4. If the final momentum of object 1 is represented by vector **H** ($p_{1,f} = H$), construct a vector diagram in the space below to find the final momentum vector, $p_{2,f}$. (Remember that $p_{1,f} + p_{2,f} = p_f$.)

5. Which vector above represents $p_{2,f}$?

   _____

HOLT PHYSICS
6 Mixed Review

Momentum and Collisions

1. A pitcher throws a softball toward home plate. The ball may be hit, sending it back toward the pitcher, or it may be caught, bringing it to a stop in the catcher's mitt.

   a. Compare the change in momentum of the ball in these two cases.

   _____

   _____

   b. Discuss the magnitude of the impulse on the ball in these two cases.

   _____

   _____

   c. In the space below, draw a vector diagram for each case, showing the initial momentum of the ball, the impulse exerted on the ball, and the resulting final momentum of the ball.

2. a. Using Newton's third law, explain why the impulse on one object in a collision is equal in magnitude but opposite in direction to the impulse on the second object.

   _____

   _____

   _____

   _____

   b. Extend your discussion of impulse and Newton's third law to the case of a bowling ball striking a set of 10 bowling pins.

   _____

   _____

   _____

   _____

Chapter

6 HOLT PHYSICS
Mixed Review *continued*

3. Starting with the conservation of total momentum, $p_f = p_i$, show that the final velocity for two objects in an inelastic collision is

$$v_f = \left(\frac{m_1}{m_1 + m_2} \right) v_{1,i} + \left(\frac{m_2}{m_1 + m_2} \right) v_{2,i}.$$

4. Two moving billiard balls, each with a mass of M, undergo an elastic collision. Immediately before the collision, ball A is moving east at 2 m/s and ball B is moving east at 4 m/s.

   a. In terms of M, what is the total momentum (magnitude and direction) immediately before the collision?

   _____

   b. The final momentum, $M(v_{A,f} + v_{B,f})$, must equal the initial momentum. If the final velocity of ball A increases to 4 m/s east because of the collision, what is the final momentum of ball B?

   _____

   c. For each ball, compare the final momentum of the ball to the initial momentum of the other ball. These results are typical of head-on elastic collisions. What generalization about head-on elastic collisions can you make?

   _____

   _____

   _____

Measuring Rotational Motion

1. Convert the following angles from degrees to radians.

   a. 17.0° _____   **c.** 50.0° _____   **e.** −20.0° _____

   b. 170.0° _____   **d.** 230.0° _____   **f.** 340.0° _____

2. Convert the following angles from radians to degrees.

   a. 1.00 rad _____   **c.** −2.50 rad _____   **e.** 3.14 rad _____

   b. 4.14 rad _____   **d.** 3.78 rad _____   **f.** 1.57 rad _____

3. A car moves forward 10.0 m in 1.5 s. Each tire rotates through an arc length of 10.0 m, and each car tire has a radius of 3.5×10^{-1} m.

   a. Find the angular displacement of one of the tires.

   b. Find the average angular speed of the tire.

   c. Assume the tire starts from rest and accelerates uniformly. Find the angular acceleration of the tire.

   d. What is the instantaneous angular speed of the tire after 1.5 s?

4. The period, *T*, of rotational motion is the time required for one complete revolution, or the time for the object to rotate through 2π rad. Starting with $\Delta\theta = \omega\Delta t$, show that $T = \dfrac{2\pi r}{v}$.

Section

HOLT PHYSICS

7-2 Concept Review

Tangential and Centripetal Acceleration

1. A wheel accelerates from rest at 1.0 rad/s^2. Find the instantaneous angular speed of the wheel at the following times.

   a. 0.10 s _____   **c.** 1.0 s _____   **e.** 5.0 s _____

   b. 0.50 s _____   **d.** 2.0 s _____   **f.** 10.0 s _____

2. If the wheel in item 1 has a radius of 0.35 m, find the tangential speed of a point on the rim of the wheel at each time in item 1.

   a. _____   **c.** _____   **e.** _____

   b. _____   **d.** _____   **f.** _____

3. If the wheel in item 1 has a radius of 0.35 m, find the tangential acceleration of a point on the rim of the wheel.

4. Find the ratio of the centripetal accelerations for the sets of rotating objects described below.

   a. $r_1 = r_2 = 2.00$ m; $v_{t,1} = 10.0$ m/s, $v_{t,2} = 5.00$ m/s

   b. $v_{t,1} = v_{t,2} = 10.0$ m/s; $r_1 = 2.00$ m, $r_2 = 1.00$ m

   c. $\omega_1 = \omega_2 = 10.0$ rad/s; $r_1 = 2.00$ m, $r_2 = 1.00$ m

5. Consider a car moving at a constant speed of 35.0 m/s on a flat road. The car turns around a curve that is 65.0 m in radius.

   a. Find the centripetal acceleration of the car. _____

   b. What provides the force necessary to make the car turn? _____

Section
7-3

HOLT PHYSICS
Concept Review
Causes of Circular Motion

1. Newton's universal law of gravitation states that $F_g = \dfrac{m_1 m_2}{r^2}$. Consider a system of two masses, $m_1 = m_2 = M$, at a distance $r = R_o$. The gravitational force on each of these masses would be $F_o = G\dfrac{MM}{R_o{}^2} = G\dfrac{M^2}{R_o{}^2}$. Find the ratio of the new gravitational force to the original force, F_o, for each of the following situations.

   a. $m_1 = M$, $m_2 = 2M$, $r = R_o$. _____

   b. $m_1 = m_2 = 2M$, $r = R_o$. _____

   c. $m_1 = m_2 = M$, $r = 2R_o$. _____

   d. $m_1 = m_2 = M$, $r = -R_o$. _____

2. For each situation in item 1, write a sentence that summarizes in words what has changed and how that change has affected the gravitational force.

   a. _____

   _____

   b. _____

   _____

   c. _____

   _____

   d. _____

   _____

3. Why is a force necessary to create circular motion?

   _____

   _____

   _____

Chapter	HOLT PHYSICS
7	# Mixed Review

Rotational Motion and the Law of Gravity

1. Complete the following table.

	s (m)	r (m)	$\Delta\theta$ (rad)	Δt (s)	ω (rad/s)	v_t (m/s)	a_c (m/s^2)
a.	4.5		1.5	0.50			
b.		0.50	8.5		8.5		
c.	3.2	0.20			58		
d.	1250		2.0	17			
e.	3750	750				86	

2. Describe the force that causes circular motion in the following cases.

a. A car exits a freeway and moves around a circular ramp to reach the street below.

_____

_____

b. The moon orbits Earth.

_____

_____

c. During gym class, a student hits a tether ball on a string.

_____

_____

3. Determine the change in gravitational force under the following changes.

a. one of the masses is doubled _____

b. both masses are doubled _____

c. the distance between masses is doubled _____

d. the distance between masses is halved _____

e. the distance between masses is tripled _____

4. Some plans for a future space station make use of rotational force to simulate gravity. In order to be effective, the centripetal acceleration at the outer rim of the station should equal about 1 g, or 9.81 m/s^2. However, humans can withstand a difference of only 1/100 g between their head and feet before they become disoriented. Assume the average human height is 2.0 m, and calculate the minimum radius for a safe, effective station. (Hint: The ratio of the centripetal acceleration of astronaut's feet to the centripetal acceleration of the astronaut's head must be at least 99/100.)

5. As an elevator begins to descend, you feel momentarily lighter. As the elevator stops, you feel momentarily heavier. Sketch the situation, and explain the sensations using the forces in your sketch.

6. Two cars start on opposite sides of a circular track. One car has a speed of 0.015 rad/s; the other car has a speed of 0.012 rad/s. If the cars start π radians apart, calculate the time it takes for the faster car to catch up with the slower car.

Section

8-1

HOLT PHYSICS

Concept Review

Torque

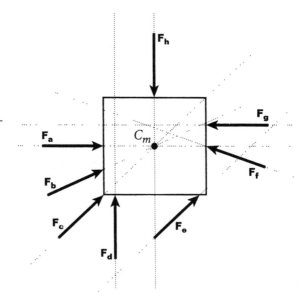

1. Use the diagram at right to complete the following items. The arrows represent force vectors, and the dashed lines represent the lines of action of the forces.

a. Identify the forces that exert a torque on the object.

_____

b. Redraw the diagram, and include only the forces that exert a torque on the object.

c. If each force has the same magnitude, which force exerts the largest torque? Explain your answer.

_____

_____

2. Two people pull on the knobs on opposite sides of a door. Sherry pulls from the inside of the door with a force of 145 N at a 90.0° angle to the door. José pulls from the outside with a 165 N force at an angle of 45.0° to the door. The doorknob is 83.0 cm from the hinge.

a. Calculate the torque Sherry exerts on the door. _____

b. Calculate the torque José exerts on the door. _____

c. Will the door rotate toward Sherry or toward José? Explain your answer.

_____

_____

HOLT PHYSICS
8-2 Diagram Skills

Rotation and Inertia

Use the diagram at right to answer items 1–4.

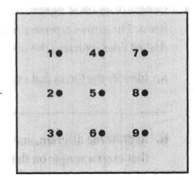

1. If the figure above has a uniform density, which point best represents the center of mass?

   _____

2. Imagine that a small hole is cut in the block at the following locations, possibly causing the center of mass to shift. In each case, identify the point toward which the center of mass will move.

   a. a single hole is cut at point 1: _____

   b. a single hole is cut at point 4: _____

   c. a single hole is cut at point 8: _____

   d. a single hole is cut at point 5: _____

3. Now imagine that a small amount of mass is added at the following locations. Again, identify the point toward which the center of mass will move.

   a. a single addition of mass is made at point 3: _____

   b. a single addition of mass is made at point 2: _____

   c. a single addition of mass is made at point 6: _____

   d. a single addition of mass is made at point 5: _____

4. If a force is applied at point 1 to the right the force will exert a clockwise torque on the object.

   a. Which two points define the lever arm for this situation? _____

   b. Where and in what direction should an equal force be applied to keep the object in equilibrium?

   _____

   _____

HOLT PHYSICS

8-3 Concept Review

Rotational Dynamics

1. A hollow ball and a solid ball have the same mass (15.0 kg) and radius (1.5 m). Both are rotating at 750 rpm.

   a. What is the angular speed of each ball?

   hollow _____ solid _____

   b. What is the moment of inertia for each ball? (Hint: Refer to **Table 8-1** on page 285 of your textbook.)

   hollow _____ solid _____

   c. What is the angular momentum of each ball?

   hollow _____ solid _____

   d. A small frictional torque of 0.10 N•m is exerted on both balls. Find the angular acceleration of each ball.

   hollow _____ solid _____

   e. Based on your answer for part d, which ball will continue to spin for a longer time?

   _____

2. A 7.3 kg bowling ball is rolled down a lane with an initial translational speed of 3.6 m/s and zero rotational speed.

   a. What is the initial energy of the ball? _____

   b. The radius of the ball is 12.0 cm. What is the moment of inertia of the ball? _____

   c. When the ball reaches the pins, it has rotational and translational kinetic energy. If the ball is rolling without slipping ($v = \omega r$), what is the translational speed of the ball? (Hint: Assume the energy from part a is conserved.) _____

   d. Frictional force makes the ball roll instead of slide. Explain how this affects the energy of the ball and how friction affects the final speed of the ball.

   _____

   _____

Section

Section 8-4

HOLT PHYSICS
Concept Review

Simple Machines

1. If friction is included in the analysis of any machine, the energy put into the machine is more than the work. How is it that simple machines make a task easier?

_____

_____

_____

2. A pulley system with a mechanical advantage of 15 is used to lift a 1750 N piano to a third-floor balcony that is 7.0 m above the ground.

   a. If friction is negligible, how much work must be done? _____

   b. What applied force must the movers use? _____

   c. How much rope will the movers pull in? _____

   d. If friction is not negligible, is the input energy greater than or less than your answer to part a?

   _____

3. Calculate the efficiency of the following.

   a. $W_{in} = 1850$ J, $W_{out} = 1700$ J _____

   b. an object weighing 150 N is lifted 9.0 m using 1500 J of energy _____

   c. a force of 150 N is exerted along a 3.0 m inclined plane to raise an object weighing 425 N to a height of 1.0 m _____

4. Explain why a real machine can never have an efficiency of 100 percent.

_____

_____

_____

5. What may be done to increase the efficiency of a real machine?

_____

_____

_____

Chapter

8 HOLT PHYSICS
Mixed Review

Rotational Equilibrium and Dynamics

1. **a.** On some doors, the doorknob is in the center of the door. What would a physicist say about the practicality of this arrangement? Why would physicists design doors with knobs farther from the hinge?

_____

_____

   b. How much more force would be required to open the door from the center rather than from the edge?

_____

2. Figure skaters commonly change the shape of their body in order to achieve spins on the ice. Explain the effects on each of the following quantities when a figure skater pulls in his or her arms.

   a. moment of inertia

_____

   b. angular momentum

_____

   c. angular speed

_____

3. For the following items, assume the objects shown are in rotational equilibrium.

   a. What is the mass of the sphere to the right?

_____

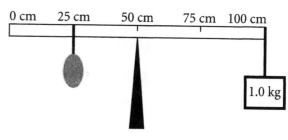

   b. What is the mass of the portion of the meter-stick to the left of the pivot? (Hint: 20% of the mass of the meterstick is on the left. How much must be on the right?)

_____

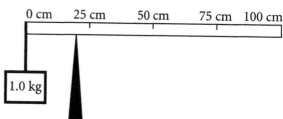

4. A force of 25 N is applied to the end of a uniform rod that is 0.50 m long and has a mass of 0.75 kg.

   a. Find the torque, moment of inertia, and angular acceleration if the rod is allowed to pivot around its center of mass. _____

   b. Find the torque, moment of inertia, and angular acceleration if the rod is allowed to pivot around the end, away from the applied force. _____

5. A satellite in orbit around Earth is initially at a constant angular speed of 7.27×10^{-5} rad/s. The mass of the satellite is 45 kg, and it has an orbital radius of 4.23×10^{7} m.

   a. Find the moment of inertia of the satellite in orbit around Earth. _____

   b. Find the angular momentum of the satellite. _____

   c. Find the rotational kinetic energy of the satellite around Earth. _____

   d. Find the tangential speed of the satellite. _____

   e. Find the translational kinetic energy of the satellite. _____

6. A series of two simple machines is used to lift a 13300 N car to a height of 3.0 m. Both machines have an efficiency of 0.90 (90 percent). Machine A moves the car, and the output of machine B is the input to machine A.

   a. How much work is done on the car? _____

   b. How much work must be done on machine A in order to achieve the amount of work done on the car? _____

   c. How much work must be done on machine B in order to achieve the amount of work from machine A? _____

   d. What is the overall efficiency of this process? _____

HOLT PHYSICS
Concept Review

Fluids and Buoyant Force

A raft is made of a plastic block with a density of 650 kg/m³, and its dimensions are 2.00 m × 3.00 m × 5.00 m.

1. What is the volume of the raft?

_____

2. What is its mass?

_____

3. What is its weight?

_____

4. What is the raft's apparent weight in water?
(Hint: density of water $= 1.00 \times 10^3$ kg/m³)

_____

5. What is the buoyant force on the raft in water?

_____

6. What is the mass of the displaced water?

_____

7. What is the volume of the displaced water?

_____

8. How much of the raft's volume is below water? How much is above?

_____

9. Answer items 5–8 using ethanol (density $= 0.806 \times 10^3$ kg/m³) instead
of water.

_____

_____

_____

_____

HOLT PHYSICS
Concept Review

Fluid Pressure and Temperature

A car's brake system transfers pressure from the main cylinder to the brake shoes on all four wheels, as shown in the diagram. The surface area of the main cylinder piston is 7.20×10^{-4} m^2 (7.20 cm^2), and that of the piston in each individual brake cylinder is 1.80×10^{-4} m^2 (1.80 cm^2). The driver exerts a 5.00 N force on the pedal.

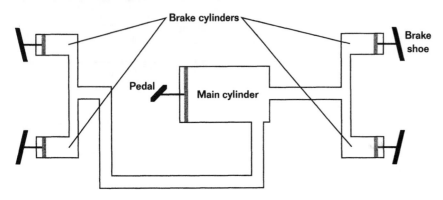

1. What is the pressure exerted on the main cylinder?

   _____

2. What is the pressure added to the liquid in this brake system?

   _____

3. What is the pressure added to each brake cylinder?

   _____

4. What is the force exerted on each brake shoe?

   _____

5. As the driver pushes the pedal, the piston moves 2.00×10^{-2} m (2.00 cm) in the main cylinder.

   a. How much volume of brake fluid is pushed out of the main cylinder?

   _____

   b. How much does the piston move in each of the brake cylinders?

   _____

Section

9-3

HOLT PHYSICS

Math Skills

Fluids in Motion

Every second, 1.20 m^3 of water enters a heating system through a pipe of medium width, *A*, with a cross-sectional area of 0.200 m^2. The water then flows into a wide pipe, *B*, with an area of 0.600 m^2, and flows out through a narrow pipe, *C*, with an area of 0.100 m^2.

1. What is the flow rate in each pipe?

   _____

2. What is the length of the segment of pipe *A* that contains 1.20 m^3 of water? Sketch the marks on the diagram above showing the segments of pipes *B* and *C* that would contain the same amount of water. What is the length of each segment?

   _____

3. How much time is required for water to travel the lengths you found in pipe *A*? in pipe *B*? in pipe *C*?

   _____

4. What is the flow speed of water in each pipe?

   _____

5. Does the speed of water increase when it enters a narrow pipe? Does the flow rate increase? Explain.

   _____

   _____

Section
9-4

HOLT PHYSICS
Concept Review

Properties of Gases

A volume of 2.40×10^{-3} m^3 of hydrogen gas is enclosed in a cylinder with a movable piston at 300 K under a pressure of 203 kPa (2.00 atm). The density of hydrogen under these conditions is 0.180 kg/m^3.

1. Calculate the mass of hydrogen in the cylinder.

_____

2. The gas is cooled down to 150 K, and the pressure is increased to 609 kPa (6.00 atm). Calculate the volume in the gas.

_____

3. What is the ratio of the final and initial temperature? pressure? volume?

_____

_____

4. How did an increase in pressure affect the volume? How did the decrease in temperature affect the volume?

_____

_____

_____

5. Did the mass of hydrogen in the cylinder increase or decrease? Explain.

_____

_____

_____

6. Find the density of hydrogen in the cylinder after the process. Has it increased or decreased? In what ratio?

_____

_____

_____

Chapter

9 HOLT PHYSICS
Mixed Review

Fluid Mechanics

1. A crate with dimensions of 2.00 m × 3.00 m × 5.00 m is immersed in sea water ($\rho = 1.025 \times 10^3$ kg/m^3) with the 3.00 × 2.00 sides as the top and bottom. The crate is held with a cable so that the top is 20.0 m below the surface of the water.

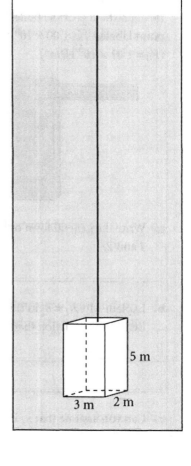

    a. Calculate the hydrostatic pressure on the top of the crate and on the bottom of the crate.

    _____

    b. Find the absolute pressure at the top and at the bottom of the crate. ($P_0 = 1.01 \times 10^5$ N/m^2)

    _____

    c. Find the forces exerted on the top and on the bottom of the crate by these pressures.

    _____

    d. On the diagram at right, sketch in vectors representing the direction and magnitude of these forces.

    e. What is the net force exerted by the water on the crate?

    _____

    f. The crate's weight is 2.50×10^6 N. Will it sink when the cable is cut? Explain.

    _____

    _____

    g. Calculate the volume of the crate.

    _____

    h. Use Archimedes' principle to find the buoyant force on the crate. How is it related to your answer to item e?

    _____

    _____

2. A very large boiler has a very small opening near the bottom, as shown in the diagram below. Water ($\rho = 1.00 \times 10^3$ kg/m^3) is constantly added through the top of the boiler to keep the boiler full. Pressure at the point labeled *1* is 1.00×10^6 N/m^2 above atmospheric pressure ($P_0 = 1.01 \times 10^5$ N/m^2).

a. Write the general form of Bernoulli's equation for the points labeled *1* and *2*.

_____

b. Explain why $h_1 = h_2$ in this case. Write the simplified form of Bernoulli's equation that results from this conclusion.

_____

_____

c. Can you assume that v_1 is approximately zero? Explain.

_____

_____

d. Write the reduced form of Bernoulli's equation that results from this assumption.

_____

e. How does P_2 compare with the atmospheric pressure P_0? How does it compare with P_1?

_____

f. Use this information to find the rate of flow of water out of the small opening. (Hint: solve Bernoulli's equation for v_2.)

_____

10-1

HOLT PHYSICS
Math Skills

Temperature and Thermal Equilibrium

1. The temperature at one of the Viking sites on Mars was found to vary daily from −90.0°F to −5.0°C. Convert these temperatures to Kelvin.

_____

2. Mercury boils at 357°C and freezes at −38.9°C.

   a. Convert these temperatures to Kelvin.

_____

   b. Can a mercury thermometer be used to measure temperatures between 500°C and 600°C? between 100°C and 200°C?

_____

_____

3. You walk out of a sauna at 45°C into a tub in which the water temperature is 309 K.

   a. Is your skin initially in thermal equilibrium with the water?

_____

   b. Is your bath going to feel cold or warm?

_____

4. Nitrogen becomes a liquid at −195.8°C under atmospheric pressure. Oxygen becomes a liquid at −183.0°C.

   a. Convert these temperatures to Kelvin.

_____

   b. A sealed tank containing a mixture of nitrogen and oxygen is cooled to 82.8 K and maintained under atmospheric pressure. Are the contents now a liquid or a gas? Explain.

_____

_____

HOLT PHYSICS

Concept Review

Defining Heat

1. A 1.000×10^3 kg car is moving at 90.0 km/hr (25.0 m/s) as it exits a freeway. The driver brakes to meet the speed limit of 36.0 km/hr (10.0 m/s).

   a. What was the car's kinetic energy on the freeway?

   _____

   b. What is its kinetic energy after slowing down?

   _____

   c. Did the internal energy of the car, road, and air increase or decrease in this process? By how much?

   _____

   d. Was work done by the car brakes and other friction forces in the process? How much?

   _____

   _____

2. A 2.00×10^2 kg sled is sliding downhill at a constant speed of 5.00 m/s until it passes a tree 20.0 m down.

   a. What was the potential energy associated with the sled and the sled's kinetic energy and total mechanical energy at the top of the hill?

   _____

   b. What were these energies at the bottom of the hill?

   _____

   c. What was the change in the sled's total energy?

   _____

   d. What was the change in the internal energy of the sled and its environment? How might that change be observed in the snow?

   _____

   _____

Section
10-3
HOLT PHYSICS
Graph Skills

Changes in Temperature and Phase

A 20.0 kg ice block is removed from a freezer whose temperature is –25.0°C and placed in an ice box with freshly caught fish. After a few hours, all the ice was melted. The final temperature of the water and the fish was 5°C.

The melting point of ice is 0.00°C. The heat capacities and latent heats are given as c_p (ice) $= 2.09 \times 10^3$ J/kg•°C; L_f (ice) $= 3.33 \times 10^5$ J/kg; c_p (water) $= 4.19 \times 10^3$ J/kg•°C. Use this information to answer the questions below.

1. How much energy did the solid ice absorb to reach its melting point and remain solid?

_____

2. How much energy was absorbed to turn the ice into water?

_____

3. How much energy was absorbed to bring the temperature of that water to 5°C?

_____

4. Draw a graph showing all of the process. (Let each box on the grid represent 0.4×10^6 J or 0.5×10^6 J.)

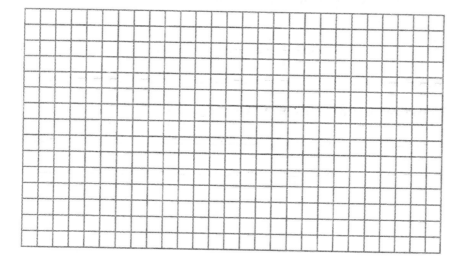

Controlling Heat

1. What is the role of the silver coating inside a thermos bottle?

   _____

   _____

   _____

2. You are cooking spaghetti atop a stove in a copper-coated stainless-steel pan filled with water. How is energy transferred from the flame to the spaghetti?

   _____

   _____

   _____

3. You are making toast for breakfast. Is most of the energy transferred from the heating element to the bread by convection or by radiation?

   _____

   _____

   _____

4. How would you answer item 3 differently if you were cooking chicken on a barbecue grille?

   _____

   _____

5. Why does wearing a wet shirt on a hot day make you feel cooler?

   _____

   _____

Chapter

10

HOLT PHYSICS
Mixed Review

Heat

1. A small bag containing 0.200 kg of lead shot at a temperature of 15.0°C falls from a 40.0 m high tower. Instead of bouncing back, the bag makes a small hole in the ground. The specific heat of lead is 1.28×10^2 J/kg•°C.

   a. Find the initial potential energy of the lead.

   _____

   b. How much energy did the lead lose by heat?

   _____

   c. The temperature of the lead after impact was 17.0°C. What was the increase in internal energy of the lead? How does it compare to the amount of lost potential energy?

   _____

   _____

   d. How much internal energy was added to the ground?

   _____

2. A very shallow pond contains 1.50×10^5 kg of water at 23°C. At the end of a windy day, 1.00×10^3 kg of water was lost by evaporation. It takes 2.26×10^6 J for 1 kg of water to evaporate.

   a. How much energy was removed from the pond by heat of evaporation?

   _____

   b. How much water was left in the pond?

   _____

   c. By how much did the temperature of the water drop in the pond? (Hint: the specific heat capacity for water is 4.19×10^3 J/(kg•°C).)

   _____

   d. Assuming there were no other changes in energy, what was the temperature of the water at the end of the day?

   _____

3. Exactly two kilograms of boiling water (100.0°C) are poured into a long, insulated aluminum pipe. The mass of the pipe is 5.000 kg, and its temperature is 20.0°C. The specific heat capacity of water is 4.19×10^3 J/kg•°C, and the specific heat capacity of aluminum is 8.99×10^2 J/kg•°C.

a. Given that the final temperature of the water is x°C and the final temperature of the pipe is y°C, explain why $y = x$.

_____

_____

b. Write expressions for the temperature change in water and in the pipe itself.

_____

c. Write an expression for the amount of energy removed from the water.

_____

d. Write an expression for the amount of energy added to the aluminum.

_____

e. Explain under what conditions these two amounts of energy may be considered equal.

_____

_____

f. Assuming that these conditions are realized, find the final temperature of the water and pipe.

_____

_____

HOLT PHYSICS

Concept Review

Relationships Between Heat and Work

1. A gas enclosed in a cylinder occupies 0.030 m^3. It is compressed under a constant pressure of 3.5×10^5 Pa until its final volume is exactly one-third of its initial volume.

   a. What was the change in the gas volume? _____

   b. How much work was done? _____

   c. The gas lost 5.0×10^3 J by heat during the compression process. Did the internal energy of the gas increase or decrease? by how much?

   _____

2. A steel marble at room temperature is placed in a plastic-foam cup containing ice and water at 0°C. After thermal equilibrium is reached, the temperature of the ice-water mixture and marble is 0°C.

   a. Was energy transferred between the marble and the water by heat? which object lost energy?

   _____

   _____

   b. Was any work done on the marble or by the marble? _____

   _____

   c. Did the internal energy of the marble increase or decrease? What was a measurable effect of this change?

   _____

   _____

   _____

   d. Did the internal energy of the water-ice mixture increase or decrease? How could this be observed?

   _____

   _____

   _____

   e. Did the internal energy of the system consisting of the water-ice mixture and the marble increase or decrease?

   _____

11-2 Diagram Skills

Thermodynamic Processes

1. A gas trapped in a cylinder does 540 J of work by expansion. At the end of the process, the internal energy has decreased by 860 J.

   a. How much energy was transferred by heat between the gas and its environment?

   _____

   b. Did the gas gain or lose energy in this transfer? Explain.

   _____

   _____

   c. In the space below, sketch a diagram of the gas container, and draw arrows showing the energy transfers by work and by heat.

2. The same amount of work (540 J) is done to **compress** the gas, this time in an **isothermal** process.

   a. What is the change in internal energy of the gas?

   _____

   b. How much energy is transferred by heat?

   _____

   c. Is that energy removed from or added to the gas? Sketch a diagram showing the energy transfers by work and by heat.

Section
11-3
HOLT PHYSICS
Concept Review

Efficiency of Heat Engines

1. A steam engine absorbs 4.00×10^4 J and expels 3.20×10^4 J by heat.

   a. How much work is done?

   _____

   b. What is the efficiency of this engine?

   _____

   c. If the engine exerts a constant force through a displacement of 25 m, how great is the force exerted by the engine?

   _____

2. The efficiency of a diesel engine is 0.35. The engine absorbs 2.00×10^4 J by heat.

   a. How much work does the engine do?

   _____

   b. How much heat is expelled?

   _____

   c. If this engine exerts a force of 175 N on an object, how far will the object be displaced?

   _____

3. An experimental gasoline engine performs at 32 percent efficiency and does 1.60×10^2 J of work in each cycle.

   a. How much energy does the engine absorb by heat in a cycle?

   _____

   b. How much energy is lost in each cycle?

   _____

   c. How much work would the same engine do if it absorbed the same amount of heat per cycle as described in **a,** but was operating at a 38 percent efficiency?

   _____

Section
11-4

HOLT PHYSICS
Math Skills

Entropy

1. A box divided by a removable partition contains two marbles in the left compartment. The partition is removed, the box is shaken, and the partition is put back into the box. Follow the steps at right to list the possible arrangements and distributions of the marbles in the box.

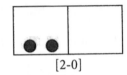

[2-0]

a. In how many ways can the marbles be arranged so that the following occur.

• both of them are in the left compartment, as in distribution [2-0]

_____

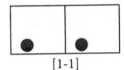

[1-1]

• each one is in different compartment, as in distribution [1-1]

_____

• both of them are in the right compartment, as in distribution [0-2]

_____

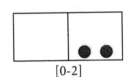

[0-2]

b. How many possible ways are there for arranging the two marbles in the box? _____

c. Which of the distributions is the most likely to occur? _____

2. Repeat the exercise above using a box that contains four marbles.

a. In how many ways can you create each of the possible distributions [4-0], [3-1], [2-2], [1-3], [0-4]?

_____

b. How many possible arrangements of the marbles are there altogether? _____

c. Which distribution is most likely to occur? _____

d. Which distribution has more disorder? _____

3. Explain how your answers about the situations of boxes with marbles relate to the increase in molecular disorder that occurs when sugar is stirred into coffee.

_____

_____

_____

Chapter
11

HOLT PHYSICS
Mixed Review

Thermodynamics

1. A system does 300 J of work at the same time that 1000 J of energy is transferred to the system by heat. What is the change in the system's internal energy?

   _____

2. Air is being compressed in a cylinder of area 0.025 m^3 under a constant pressure of 3.0×10^5 Pa, and the volume of the air in the cylinder is reduced to 0.020 m^3.

   a. By how much is the volume of air reduced?

   _____

   b. How much work is done in the process?

   _____

   c. The cylinder is thermally insulated, making the process adiabatic. What is the change in internal energy of the gas?

   _____

3. A gasoline engine runs with 28 percent efficiency. It expels 3.60×10^4 J of heat in each cycle.

   a. Find the heat absorbed in one cycle.

   _____

   b. Find the work output in one cycle.

   _____

4. When you use a pump to push air into a bicycle tire, the pump and the air eventually warm up.

   a. Explain how this is related to the first law of thermodynamics.

   _____

   _____

   b. Explain how this fact is related to the second law of thermodynamics.

   _____

   _____

5. A basketball bounces to half of its original height when dropped. In the space below, sketch energy bar diagrams describing the ball's potential energy, the ball's kinetic energy, the internal energy of the ball, and the ball's environment at each of the following four instants.

- just before the ball is dropped

- immediately after the first bounce

- at its highest point after the first bounce

- immediately after the second bounce

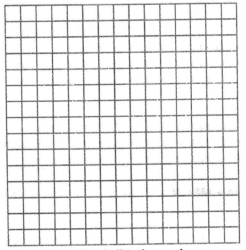

Before ball is dropped

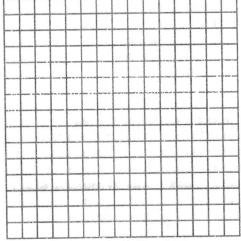

Immediately after first bounce

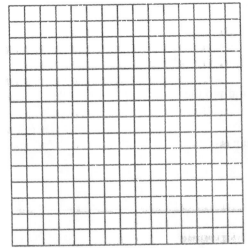

At high point after first bounce

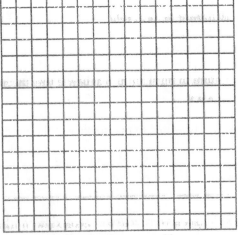

Immediately after second bounce

12-1

HOLT PHYSICS

Concept Review

Simple Harmonic Motion

1. A clown is rocking on a rocking chair in the dark. His glowing red nose moves back and forth a distance of 0.42 m exactly 30 times a minute, in a simple harmonic motion.

    a. What is the amplitude of this motion?

    _____

    b. What is the period of this motion?

    _____

    c. What is the frequency of this motion?

    _____

    d. The top of the clown's hat contains a small light bulb that shines a narrow light beam. The beam makes a spot on the wall that goes back and forth between two dots placed 1.00 m apart as the clown rocks. What are the amplitude, period, and frequency of the spot's motion?

    _____

    _____

    _____

2. A 5.00 kg block hung on a spring causes a 10.0 cm elongation of the spring.

    a. What is the restoring force exerted on the block by the spring?

    _____

    b. What is the spring constant?

    _____

    c. What force is required to stretch this spring 8.50 cm horizontally?

    _____

    d. What will the spring's elongation be when pulled by a force of 77.7 N?

    _____

HOLT PHYSICS
Math Skills

Measuring Simple Harmonic Motion

1. A spring-mass system vibrates exactly 10 times per second. Find its period and its frequency.

   _____

2. A pendulum swings with a period of 0.20 seconds.

   a. What is its frequency?

   _____

   b. How many times does it pass the lowest point on its path in 1.0 second? in 7.0 seconds?

   _____

3. A spring-mass system completes 20.0 vibrations in 5.0 seconds, with a 2.0 cm amplitude.

   a. Find its frequency and its period.

   _____

   b. The same mass is pulled 5.0 cm away from the equilibrium position, then released. What will the period, the frequency, and the amplitude be?

   _____

4. A pendulum completes 30.0 oscillations per minute. Find its frequency, its period, and its length.

   _____

5. A spring has a 2.000×10^3 N/m spring constant.

   a. What mass will make it oscillate 5.0 times per second? 10.0 times per second?

   _____

   b. You want the mass-spring system to operate at a higher frequency. Should you increase or decrease the mass?

   _____

Properties of Waves

1. Radio waves travel at the speed of light (3.00×10^8 m/s). An amateur radio system can receive radio signals at frequencies between 8.00 MHz and 1.20 MHz. What is the range of the wavelengths this system can receive?

_____

_____

_____

2. Graph **(a)** below describes the density versus time of a pressure wave traveling through an elastic medium. Graph **(b)** describes the density versus distance for the same wave.

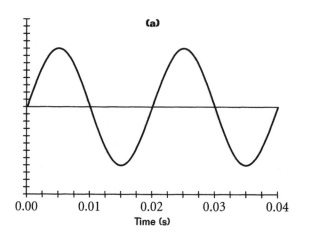

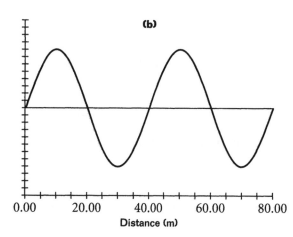

a. Use graph **(a)** to find the period of oscillation of this wave and its frequency.

_____

b. Use graph **(b)** to find the wavelength and the speed.

_____

Section

12-4

HOLT PHYSICS

Graph Skills

Wave Interactions

1. A wave of 0.25 cm amplitude traveling on a string interferes with a wave of 0.35 cm amplitude that was generated at the other end with the same frequency. Their maxima occur at the same points on the string.

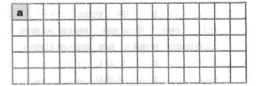

  a. Sketch a graph of each individual wave traveling through the same area of the string for one period on the grids labeled (**a**) and (**b**).

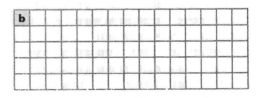

  b. Sketch a graph of the wave shape resulting from inter-ference on the grid labeled (**c**).

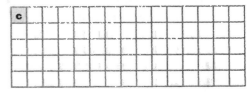

2. A 15.0 m long string is tied at one end (point *B*) and shaken repeatedly at the other end (point *A*) with a 2.00 Hz frequency. This generates waves that travel at 20.0 m/s in the string.

  a. How long does it take for each pulse to travel from *A* to *B* and return to *A*?

  _____

  b. What is the wavelength of these waves?

  _____

  c. Are the pulses inverted when reflected from *B*?

  _____

HOLT PHYSICS
12 Mixed Review

Vibrations and Waves

1. A pendulum with a mass of 0.100 kg was released. The string made a 7.0° angle with the vertical. The bob of the pendulum returns to its lowest point every 0.10 s.

   a. What is its period? What is its frequency?

   _____

   _____

   b. The pendulum is replaced by one with a mass of 0.300 kg and set to swing with a 15° angle. Do the following quantities increase, decrease, or remain the same?

   period _____

   frequency _____

   total energy _____

   speed at the lowest point _____

2. A narrow, flat steel rod is anchored at its lower end, with a 0.500 kg ball welded to the top end. A force of 6.00 N is required to hold the ball 10.0 cm away from its central position.

   If this arrangement is modeled as an oscillating horizontal mass-spring system, vibrating with a simple harmonic motion, find

   a. the force constant, k, of the spring.

   _____

   b. the period and frequency of the oscillations.

   _____

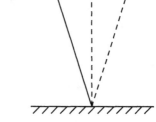

3. Find the acceleration due to gravity at a place where a simple pendulum 0.150 m long completes 1.00×10^2 oscillations in 3.00×10^2 seconds. Could this place be on Earth?

   _____

   _____

4. Consider the first two cycles of a pendulum swinging from position A with a period of 2.00 s.

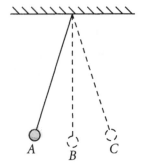

   a. At which times is the bob found at positions A, B, and C during the first two cycles?

   _____

   _____

   b. At which times and locations is gravitational potential energy at a maximum? At which times is kinetic energy at a maximum?

   _____

   _____

   c. At which times and locations is the velocity at a maximum? the restoring force? the acceleration?

   _____

   _____

   _____

5. The frequency of a pressure wave is 1.00×10^2 Hz. Its wavelength is 3.00 m. Find the speed of wave propagation.

   _____

6. A pressure wave of 0.50 m wavelength propagates through a 3.00 m long coil spring at a speed of 2.00 m/s. How long does it take for the wave to travel from one end of the coil to the other? How many wavelengths fit in the coil?

   _____

   _____

13-1

HOLT PHYSICS

Concept Review

Sound Waves

1. In an experiment for measuring the speed of sound, a gun was shot 715 m away from the observer. It was heard 2.13 seconds after the flash was seen. What was the speed of sound in air at that time?

   _____

2. Sound travels at 1530 m/s in sea water. A signal sent down from a ship is reflected at the bottom of the ocean and returns 1.35 s later. Assuming the speed of sound was not affected by changes in the water, how deep was the ocean at that point?

   _____

3. A train at rest blows a whistle to alert passengers that it is about to depart from a subway station. The pitch of this whistle is 1.14×10^4 Hz. The speed of sound in the air in that subway tunnel is 342 m/s. The speed of sound in iron is 5130 m/s.

   a. What is the wavelength of that sound in the air?

   _____

   b. What is the distance between consecutive areas of compression and of rarefaction in the spherical sound waves spreading from the whistle in the air?

   _____

   c. Assuming that the sound was loud enough to be heard from the end of the 1200 m long tunnel, when was it heard through air? through the rails?

   _____

   d. What was the apparent frequency of the sound waves that reached the end of the tunnel?

   _____

   e. As the train left the station, did the frequency appear to change for a listener on the platform? inside the train? at the other end of the tunnel?

   _____

   _____

HOLT PHYSICS
Concept Review

Sound Intensity and Resonance

Refer to the following table to answer the following questions.

Intensity (W/m^2)	Decibel level (dB)	Intensity (W/m^2)	Decibel level (dB)
1.0×10^{-9}	30	1.0×10^{-5}	70
1.0×10^{-8}	40	1.0×10^{-4}	80
1.0×10^{-7}	50	1.0×10^{-3}	90
1.0×10^{-6}	60	1.0×10^{-2}	100

1. While practicing his instrument at home, a young drummer produces sounds with 0.5 W of power. Assume the sound waves spread spherically, with no absorption in the medium.

a. What is the intensity of the sound waves that reach the walls of his room 2.00 to 4.00 m from the drum?

_____

b. What is the intensity of the sound waves that reach the family room, 8.00 to 12.0 m from the drum?

_____

c. What is the intensity and approximate decibel level of the sound waves that reach the neighbors' home 50.0 m away?

_____

2. The sound level 5.00 meters away from a jackhammer is exactly 100 dB.

a. What is the intensity of the sound at that point?

_____

b. What is the power of the sound from the jackhammer?

_____

c. At what distance from the jackhammer will the noise intensity decrease to 1.00×10^{-8} W/m^2?

_____

Section
13-3
HOLT PHYSICS
Diagram Skills

Harmonics

1. A 52.0 cm long guitar string has a fundamental frequency of 444 Hz.

   a. What is the speed of sound in the string according to these data?

   _____

   b. In the space below, draw the standing wave pattern for the first, the second, and the third harmonics, showing the nodes and the antinodes on the string.

   c. What should be the string's length in order to produce a fundamental note of 333 Hz?

   _____

2. The first harmonic frequency of a violin string is 440 Hz.

   a. Find the next harmonic frequencies (overtones) of this string.

   _____

   b. The intensities of the second and third harmonics are about half that of the fundamental one. Sketch a graph of each wave and a graph of their combination to show the resultant waveform for this violin string.

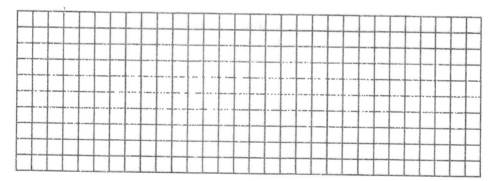

HOLT PHYSICS
13 Mixed Review

Sound

1. The speed of sound increases with temperature. It is 331 m/s in air at 0°C and 343 m/s in air at 20°C. A glass pipe vibrates with a frequency of 151 Hz.

   a. What is the wavelength of the sound produced by the column of air in the pipe on a cold day (0°C) and on a warmer day (20°C)?

   _____

   _____

   b. How does air temperature affect the wavelength of the sound produced by the pipe?

   _____

   _____

2. The driver of an ambulance turns on its siren as the ambulance heads east at 30 mph. A police car is following the ambulance at 30 mph. A truck behind the police car is moving at 20 mph. A van is traveling west in the opposite lane at 20 mph. A small car is stopped at the side of the road. The vehicles are positioned as shown.

   a. On the diagram, sketch and label arrows to indicate the velocity of each vehicle.

   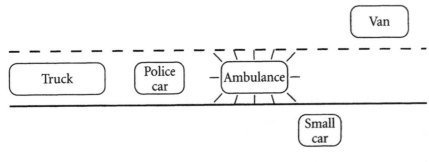

   b. Rank the sounds perceived by the passengers in each of the vehicles in order of decreasing frequency.

   _____

   _____

   _____

   _____

3. A 330 Hz tuning fork is vibrating after being struck. It is placed on a table near but not directly touching other objects, including other tuning forks. Eventually one glass and one other tuning fork start vibrating. Explain why this happens.

_____

_____

_____

_____

4. The first harmonic in a pipe closed at one end is 487 Hz.

   a. Find the next two harmonic frequencies that will occur in this pipe.

_____

   b. What are the corresponding wavelengths of the first three harmonics? (Hint: assume the speed of sound is 345 m/s.)

_____

   c. What is the length of this pipe?

_____

   d. Repeat this exercise for a pipe open at both ends.

_____

_____

_____

_____

5. A piano tuner uses a 440 Hz tuning fork to tune a string that is currently vibrating at 445 Hz.

   a. How many beats per second does he hear?

_____

   b. What other frequency could produce the same sound effect? Explain why.

_____

Section
14-1

HOLT PHYSICS
Concept Review

Characteristics of Light

1. The orbital radius of the Earth (the average Earth-Sun distance) is 1.496×10^{11} m. Mercury's orbital radius is 5.79×10^{10} m and Pluto's is 5.91×10^{12} m. Calculate the time required for light to travel from the Sun to each of the three planets. (Hint: Use 3.00×10^8 m/s for the speed of light.)

   a. Sun-Earth _____

   b. Sun-Mercury _____

   c. Sun-Pluto _____

2. Typical wavelengths of visible light colors are listed below.

violet	blue	green	orange-yellow	red
420 nm	450 nm	550 nm	600 nm	700 nm

   a. Calculate the frequency of the electromagnetic waves that carry these colors.

   _____

   _____

   b. How does frequency change when wavelength increases?

   _____

   _____

   _____

   _____

   c. Does the speed of light in air depend on frequency? on wavelength?

   _____

   _____

   _____

   _____

Section
14-2
HOLT PHYSICS
Diagram Skills

Flat Mirrors

1. The point of a 20.0 cm pencil is placed 25.0 cm from a flat mirror. Its eraser is 15.0 cm from the mirror. Three of the light rays from the pencil's point hit the mirror with incident angles of 0°, 20°, and 50° at points *A*, *B*, and *C* as shown.

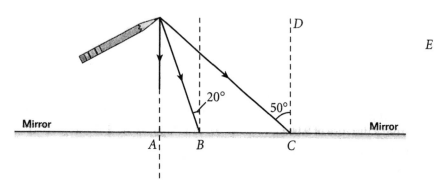

a. Use a protractor to draw the reflected rays from points *A*, *B*, and *C*.

b. Where do reflected rays or their extensions intersect?

_____

c. What is the distance between the pencil's head and its image?

_____

d. Would a person's eye located at point *D* perceive one of the reflected rays you drew? Will the person be able to see the image? Explain.

_____

_____

e. What if the eye is located at point *E*?

_____

f. Draw incident rays from the eraser of the pencil to point *A* and to point *B*. Measure their incident angles and write them on the line below.

_____

g. Draw the reflected rays and locate the image of the eraser. Draw the pencil's image.

HRW material copyrighted under notice appearing earlier in this book.

Section

14-3
HOLT PHYSICS
Diagram Skills

Curved Mirrors

1. A 1.50 m tall child is in a mirror gallery at the amusement park. She is standing in front of a concave mirror with a radius of 4.00 m. She starts walking toward the mirror from a distance of 9.00 m, and she stops every meter to observe her image.

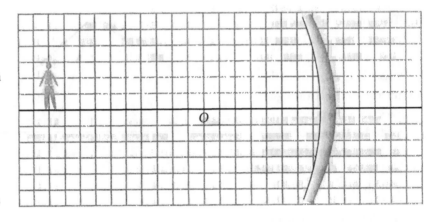

   a. Find the focal point of this mirror and label it *F*.

   b. Mark the child's locations 9.00 m, 5.00 m, and 1.00 m in front of the mirror, and label them *A, B, C*.

   c. Sketch ray diagrams to locate the image formed when the child is at *A*. Measure the distance from the image to the mirror and record it below.

      Distance of *A*'s image = _____

   d. Repeat question c for the object at positions *B* and *C*.

      Distance of *B*'s image = _____

      Distance of *C*'s image = _____

2. Calculate the image location for the object at *A, B*, and *C* in item 1, using the mirror equation. Compare your results with your diagrams.

      Distance of *A*'s image = _____

      Distance of *B*'s image = _____

      Distance of *C*'s image = _____

HOLT PHYSICS

Section 14-4 Concept Review

Color and Polarization

1. It is common knowledge that chlorophyll allows green plants to use light for photosynthesis.

    a. Which colors of the visible spectrum do green plants absorb? Explain.

_____

_____

_____

    b. A window has just broken in your greenhouse. Until it can be replaced, you can seal the hole with clear plastic that is slightly tinted either red or green. Which would you use? Explain.

_____

_____

_____

2. You have three spotlights: one red, one green, one blue. You also have three buckets: one with red paint, one with green paint, one with blue paint.

    a. What color do you see when you shine all three spotlights on a white wall in a dark room?

_____

    b. What color do you see if you paint the wall blue before shining all three spotlights on it in a dark room?

_____

    c. What color do you see when you paint the wall with a brush dipped in the red and blue buckets and then shine green light on it?

_____

    d. What color do you see when you paint the wall with a brush dipped in all three buckets and then shine all three spotlights on it?

_____

3. What color do you see when shining green light on a magenta painting?

_____

| Chapter | **HOLT PHYSICS** |
| 14 | # Mixed Review |

Light and Reflection

1. Proxima Centauri, the nearest star in our galaxy, is 4.30 light-years away. What is its distance in meters?

_____

2. Radio signals emitted from and received by an airplane have a frequency of 3.00×10^{12} Hz and travel at the speed of light.

  a. How long is the delay in each message going from the control tower to a jet flying at 1.00×10^4 m of altitude?

_____

  b. What is the wavelength of these signals?

_____

3. A laser beam is sent to the moon from Earth. The reflected beam is received on Earth after 2.56 seconds. What is the distance from the moon the Earth?

_____

4. The background radiation in the universe (believed to come from the Big Bang) includes microwaves with wavelengths of 0.100 cm. What is the frequency of this radiation?

_____

5. List five objects that reflect light diffusely. List three objects that reflect light specularly for the most part.

Diffuse reflection _____

_____

Specular reflection _____

_____

Chapter

14 HOLT PHYSICS
Mixed Review *continued*

6. A mirror door is located next to a large wall mirror.
The door is closed to create a 90° angle with the
wall. You stand 2.00 m from the door and 1.00 m
from the wall.

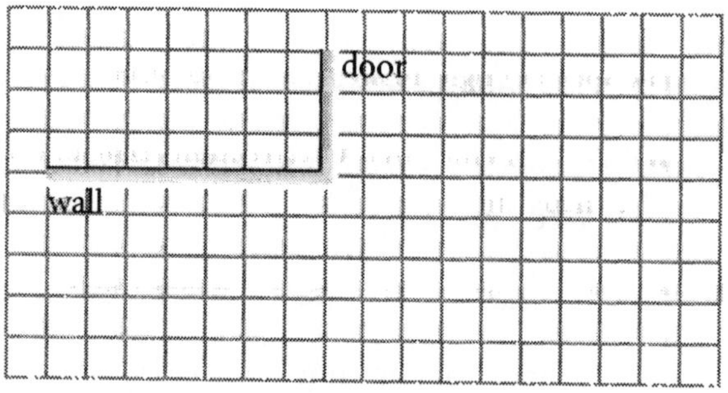

a. On the diagram at right, sketch a top-view diagram of the situation at scale. Label the object
(yourself) *A*.

b. Locate your first image in the mirror on the
door. Label it *B*. Locate *B*'s image in the mirror
on the wall. Label it *C*.

c. Locate your first image in the mirror on the wall and its image in the
mirror on the door. Label them *D* and *E*.

d. Where will the next images of the images be located?

_____

7. An object located 36.0 cm from a concave mirror produces a real image
located 12.0 cm from the mirror.

a. Find the focal length of this mirror

_____

b. Find the location, type, and size of the image formed by a 6.00 cm tall
object located 30.0 cm, 24.0 cm, 18.0 cm, 12.0 cm, and 6.00 cm in
front of the mirror.

_____

_____

_____

8. The concave mirror in the problem above is replaced by a convex one
with the same curvature. Find the location of the images produced when
the object is located 30.0 cm, 24.0 cm, 18.0 cm, 12.0 cm, and 6.00 cm in
front of the mirror.

_____

_____

_____

HOLT PHYSICS
Concept Review

Refraction

1. The speed of light in air is 3.00×10^8 m/s.

   a. How does the index of refraction relate to the speed of light in a medium?

   _____

   b. The index of refraction of water is 1.33. What is the speed of light in water?

   _____

2. A light ray traveling in air strikes a glass plate with a refractive index of 1.52 at a 20.0° angle from the normal. After refraction, going in and out of the glass, the exiting ray forms an angle θ with the normal to the surface on the other side.

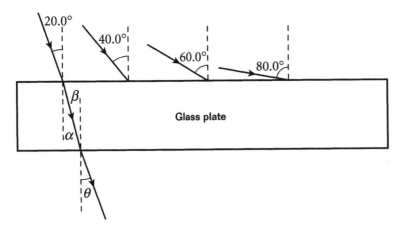

   a. Find α, the angle of refraction from air to glass. _____

   b. The plate sides are parallel. Find β, the angle of incidence from glass to air, and θ, the angle of refraction.

   _____

   c. Repeat when the angle of incidence from air is 40°, 60°, and 80°.

   _____

   _____

   d. Sketch the results on the diagram above.

HOLT PHYSICS
Diagram Skills

Thin Lenses

1. A converging lens has a focal length of 3.00 cm. The letters *A*, *B*, and *C* are used as objects placed at distances of 8.00 cm, 5.00 cm, and 2.00 cm, respectively, from the lens.

   a. Sketch ray diagrams to locate the image of *A*: Draw one ray from the top of the head parallel to the axis and another ray from the head through the focal point. Verify that the image is also in the ray that passes through the center of the lens.

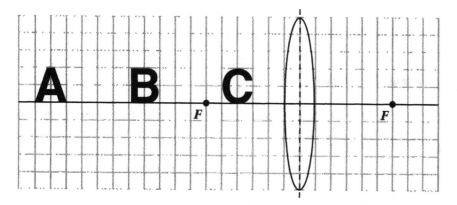

   b. Is the image of *A* real? inverted? magnified?

   _____

   c. Repeat questions a and b for the object at positions *B* and *C*.

   _____

   _____

2. Calculate the image location for the object at *A*, *B*, and *C* in problem 1. Compare your results with your diagrams.

   _____

   _____

   _____

HOLT PHYSICS
Concept Review

Optical Phenomena

Indices of Refraction for Various Substances

Substance	n
Diamond	2.419
Sodium chloride	1.544
Glycerine	1.473
Fluorite	1.434

Substance	n
Ethyl alcohol	1.361
Water	1.333
Air	1.000

1. A light ray inside a diamond strikes the boundary with air at 20.0° from the normal.

   a. Calculate the angle of refraction of that light ray.

   _____

   b. What happens when the incident angle is 32.0°?

   _____

   c. What is the critical angle for this light traveling from diamond to air?

   _____

   d. The diamond is immersed in water. The same light ray strikes the diamond-water boundary at a 20.0° angle. Answer items a, b, and c for this case.

   _____

   _____

2. Glass prisms with 90°, 45°, 45° angles are used in periscopes because light entering the right-angle side undergoes internal reflection on the 45° side of the prisms. What happens if the sides of the prisms are made of thin glass and the prisms are filled with water? Use the critical angle of water to answer.

   _____

   _____

   _____

HOLT PHYSICS
Mixed Review

Refraction

1. Two parallel rays enter an aquarium as shown. Ray 1 forms a 70.0° angle with the normal to the surface. Ray 2 forms a 20.0° angle with the normal to the wall. (Hint: the index of refraction for water is 1.33.)

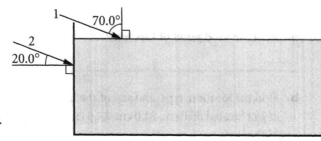

a. Calculate the angle of refraction of each ray.

_____

b. Trace the path of each light ray inside the water.

c. Are the refracted rays inside the water still parallel? Will they intersect in the water?

_____

_____

2. A large beaker contains layers of water of increasing salinity, separated by a thin plastic plate. The lowest layer has the highest salinity and refractive index, as shown in the diagram. A ray of light strikes the surface of fresh water at the top, at a 70.0° angle from the normal.

air	70.0°	$n = 1.00$
fresh water		$n = 1.33$
salt water		$n = 1.45$
high salinity		$n = 1.57$

a. Find the angles of refraction and the angles of incidence at each boundary.

_____

_____

_____

_____

b. There is a flat mirror at the bottom of the container. Trace the path of one light ray coming from the air to the bottom of the beaker and back.

Chapter

HOLT PHYSICS

15 Mixed Review *continued*

3. An object located 36.0 cm from a thin converging lens has a real image located 12.0 cm from the lens.

a. Find the focal point of this lens.

_____

b. Find the location, type, and size of the image formed by a 6.00 cm tall object located 30.0 cm, 24.0 cm, 18.0 cm, 12.0 cm, and 6.00 cm in front of the lens.

_____

_____

_____

_____

_____

4. The converging lens in item 3 is replaced by a diverging lens. Now the image of the first object is located 12.0 cm in front of the lens. Find the focal distance of the diverging lens and the location of the images produced when the object is placed at the distances described in item 3b.

_____

_____

_____

_____

_____

5. A bug placed 1.00 cm under a magnifying glass appears exactly six times larger.

a. Where is the bug's image located?

_____

b. What is the focal point of the lens in the magnifying glass?

_____

HOLT PHYSICS

16-1 Concept Review

Interference

1. Monochromatic light with a wavelength of 560 nm is used in a double-slit experiment. The distance between the slits was 2.00×10^{-5} m.

   a. Find the angle of the first, second, and third bright fringes on the screen.

   _____

   _____

   b. The experiment is repeated with the distance between slits at 2.00×10^{-6} m. Find the angles of the first three bright fringes.

   _____

   _____

   c. How does the separation between fringes change when the distance between slits changes? What would you observe if the distance between slits is 2.00 cm?

   _____

   _____

   _____

2. The distance between two slits in a double-slit experiment is 7.00×10^{-6} m. The first order bright fringe produced by monochromatic light appears on the screen at an angle of 3.89° from the central maximum.

   a. Determine the wavelength of light used in this experiment.

   _____

   b. Find the angles of the second, third, and fourth bright fringes.

   _____

   _____

Section
16-2

HOLT PHYSICS
Concept Review

Diffraction

1. A diffraction grating has 8.00×10^3 lines per centimeter.

   a. What is the slit spacing in this grating?

   _____

   b. Is the grating appropriate for observing the diffraction of visible light
   (400 to 700 nm)? For better results, would you choose a grating with
   wider spacing? with more lines per centimeter? Explain.

   _____

   _____

   _____

2. The spacing in a diffraction grating is 8.00×10^{-6} m.

   a. How many lines per centimeter are there?

   _____

   b. Find the first, second, and third angles at which one would observe
   maxima when light of 620 nm wavelength is diffracted.

   _____

   _____

3. The second-order maxima are observed at 8.12° with the grating above
   in a diffraction experiment. What is the wavelength?

   _____

4. Monochromatic light of 570 nm is diffracted by a grating of unknown
   spacing. The third-order maxima are observed at a 23° angle. What is the
   spacing in that grating?

   _____

HOLT PHYSICS
Concept Review

Coherence

1. Describe the term *coherent light.*

_____

_____

_____

2. Draw a diagram that illustrates coherent light and noncoherent light.

3. What type of energy is used to cause the stimulated emission?

_____

_____

4. List three applications of lasers.

_____

_____

Chapter	**HOLT PHYSICS**
16	**Mixed Review**

Interference and Diffraction

1. The second-order bright fringes of interference are observed at an 8.53°
   angle in a double-slit experiment with light of 5.00×10^2 nm wavelength.

   a. Determine the slits' separation.

   _____

   b. Find the angle of the tenth-order bright fringe.

   _____

   c. In this experiment, the screen is 2.00 m wide. Its distance from the
   source is 1.00 m. Could the tenth-order fringe be observed? Why or
   why not?

   _____

   _____

   _____

   _____

2. Diffraction of white light with a single slit produces bright lines of
   different colors.

   a. Which wavelengths are more diffracted by the same slit size?

   _____

   _____

   b. In the space below, sketch a diagram showing the location of red, green
   and blue lines of the first and second order. Describe the sequence in
   which the colors appear, beginning with the color closest to the center.

   _____

   _____

   c. What is the color of the central image?

   _____

3. You have three diffraction gratings. Grating A has 2.0×10^5 lines per meter. Grating B has 9.0×10^6 lines per meter. Grating C has 3.0×10^7 lines per meter.

a. What is the slit distance of each grating?

_____

_____

b. Which gratings can diffract the following:

• visible light of 500 nm wavelength

_____

• X rays of 5.00 nm wavelength

_____

• infrared light of 5000 nm wavelength

_____

4. You drop pebbles into the water on a rocky beach. When the waves you made reach the rocks, new waves appear to start in the spaces between the rocks.

a. Are these waves coherent?

_____

_____

b. How is this like a double slit illuminated by a single light source?

_____

_____

_____

17-1

HOLT PHYSICS

Concept Review

Electric Charge

1. A plastic rod rubbed with wool was used to charge a small metal sphere in three experiments, as illustrated below. The spheres were held by insulating stands. The sphere in Experiment B was grounded. Assume the rod had a positive charge.

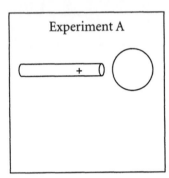
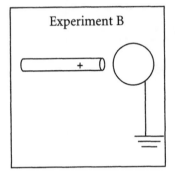
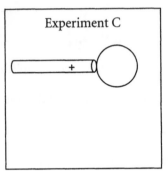

   a. Were charges transferred in Experiments A, B, or C? If so, between which objects?

   _____

   _____

   _____

   b. Sketch the charge distribution for the spheres in each experiment.

   c. The rod was removed after a while. In which experiment(s) did the sphere end up with excess electric charge?

   _____

   d. In which experiment(s) did polarization occur?

   _____

   e. What happened to the excess charge on the rod after it was removed in experiment A? in B? in C?

   _____

   _____

   _____

Section
17-2
HOLT PHYSICS
Math Skills

Electric Force

Use $k_C = 8.99 \times 10^9 \text{ N} \cdot \text{m}^2/\text{C}^2$.

1. Two point charges, q_1 and q_2, of 4.00 μC each, are placed -16.0 cm and 16.0 cm away from the origin on the x-axis. A charge q_3 of -1.00 μC is placed 12.0 cm away from the origin on the y-axis.

   a. Find the distance from q_3 to q_1 and from q_3 to q_2 _____

   b. Find the magnitude and the direction of the force F_{13} exerted by q_1 on q_3. _____

   c. Find the magnitude and the direction of the force F_{23} exerted by q_2 on q_3. _____

   d. Find the magnitude and the direction of the force F_{12} exerted by q_1 on q_2. _____

   e. In the space below, sketch the vectors representing forces F_{13} and F_{23}.

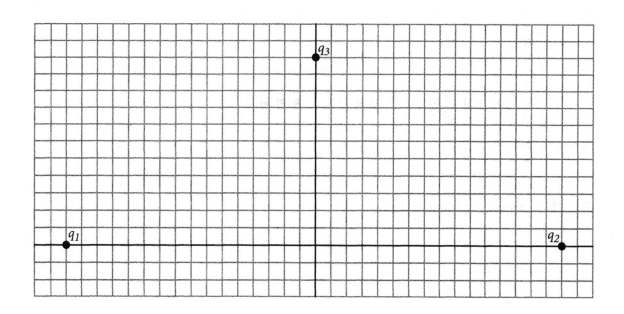

   f. Find the angle between the q_1–q_3 line and the x-axis. _____

   g. Find the x and y components of forces F_{13} and F_{23}. _____

   h. Find the resultant force of forces F_{13} and F_{23}. _____

   i. If q_3 is released, in which direction will it move? _____

Section

17-3

HOLT PHYSICS

Concept Review

The Electric Field

Use $k_C = 8.99 \times 10^9 \, \text{N} \cdot \text{m}^2/\text{C}^2$.

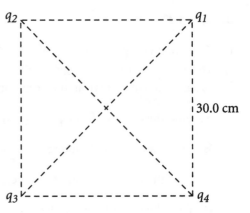

1. Four positive charges, q_1, q_2, q_3, and q_4, of 8.00 μC, each are arranged to form a 30.0 cm wide square as shown.

   a. Find the distance of each charge from the center of the square.

   _____

   b. Find the strength and direction of the electric field vectors of q_1, q_2, q_3, and q_4 at the center of the square.

   _____

   _____

   c. Find the strength and direction of the electric field at the center of the square.

   _____

2. In a Millikan experiment, a droplet of mass 4.7×10^{-15} kg floats in an electric field of 3.20×10^4 N/C.

   a. What is the force of gravity on this droplet?

   _____

   b. What is the electric force that balances it?

   _____

   c. What is the excess charge?

   _____

   d. How many excess electrons are there on this droplet?

   _____

Chapter

17 | HOLT PHYSICS
Mixed Review

Electric Forces and Fields

Use $k_C = 8.99 \times 10^9 \, \text{N} \cdot \text{m}^2/\text{C}^2$.

1. Two spheres, *A* and *B*, are placed 0.60 m apart, as shown. Sphere *A* has +3.00 μC excess charge. Sphere *B* has +5.00 μC excess charge.

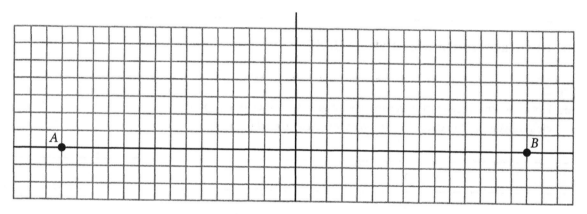

   a. How many electrons are missing on sphere *A*? on sphere *B*?

   _____

   b. How do the forces of *B* on *A* and *A* on *B* compare? Does the greater charge exert a greater force?

   _____

   _____

2. A third spherical charge, *C*, of +2.00 μC, is placed on the line connecting spheres *A* and *B*. Find the resultant force exerted by *A* and *B* on *C* when *C* is placed in the following locations.

   a. 0.20 m to the left of *A*

   _____

   b. 0.20 m to the right of *A* between *A* and *B*

   _____

   c. exactly in the middle between *A* and *B*

   _____

3. Alpha particles are made of two protons and two neutrons.
$m_p = 1.673 \times 10^{-27}$ kg; $m_n = 1.675 \times 10^{-27}$ kg; $q_e = 1.60 \times 10^{-19}$ C

  a. Find the electric force acting on an alpha particle in a horizontal
  electric field of 6.00×10^2 N/C.

  _____

  b. What is the acceleration of this alpha particle?

  _____

  c. How does this acceleration compare with gravity? Describe the particle's trajectory. Will it be close to horizontal? to vertical free fall?

  _____

  _____

  _____

  _____

4. A 2.00 μC point charge of mass 5.00 g is suspended on a string and
placed in a horizontal electric field. The mass is in equilibrium when the
string forms a 17.3° angle with the vertical.

  a. In the space below, sketch a free-body diagram of the problem. Show the
  vertical and horizontal components of the tension force in the string.

  b. Find the electric force on the charge in this field.

  _____

  c. Find the strength of the electric field.

  _____

5. How many electrons are there in 1.00 C? How many electrons are there
in 1.00 μC?

  _____

  _____

HOLT PHYSICS
Concept Review

Electrical Potential Energy

Use $k_C = 8.99 \times 10^9$ N•m^2/C^2.

1. A positive charge, q_1, of 5.00×10^{-9} C is placed at $(-20.0$ cm, 0$)$ of a coordinate system. An equal and opposite charge, q_2, is at $(20.0$ cm, 0$)$. Sketch a diagram for each of the questions below.

   a. What is the potential energy of this pair of charges? Was work done to bring q_2 from infinity to its place near q_1? How much?

   _____

   _____

   b. A positive charge, q_3, equal to q_1 is placed at $(60.0$ cm, 0$)$. What is the potential energy of the three charges? Was work done on or by the charges for bringing q_3 from infinity to its place near q_1 and q_2? How much?

   _____

   _____

2. An alpha particle travels 5.00 cm in a uniform electric field of 6.00×10^2 N/C. (Alpha particles are made of two protons and two neutrons. $m_p = 1.673 \times 10^{-27}$ kg; $m_n = 1.675 \times 10^{-27}$ kg; $q_e = 1.60 \times 10^{-19}$ C)

   a. What is the change in the potential energy of the particle? Does it increase or decrease?

   _____

   b. If the particle is initially at rest, what is its final kinetic energy?

   _____

   c. What is its speed?

   _____

HOLT PHYSICS

Section

18-2 | Concept Review

Potential Difference

1. A point charge, q_1, of 2.00 μC is placed on the x-axis at (-4.00 cm, 0 cm).
An identical charge, q_2, is placed at (4.00 cm, 0 cm). Find the total elec-
tric potential due to these charges at the following locations. Use
$k_C = 8.99 \times 10^9$ N•m^2/C^2.

a. the center $(0, 0)$ _____

b. on the y-axis at

• $y = -10.0$ cm _____

• $y = -2.00$ cm _____

• $y = 2.00$ cm _____

• $y = 10.0$ cm _____

c. on the x-axis at

• $x = -10.0$ cm _____

• $x = -2.00$ cm _____

• $x = 2.00$ cm _____

• $x = 10.0$ cm _____

2. Find the electric potential at the center of a square with four point charges
$q_1, q_2, q_3, q_4,$ placed at (5.00 cm, 0 cm), (0 cm, 5.00 cm), (-5.00 cm, 0 cm),
and (0 cm, -5.00 cm), respectively, for the following cases.

a. $q_1 = q_2 = q_3 = q_4 = 3.00\ \mu C$

_____

b. $q_1 = q_3 = 3.00\ \mu C; q_2 = q_4 = -3.00\ \mu C$

_____

c. $q_1 = q_2 = 3.00\ \mu C; q_3 = q_4 = -3.00\ \mu C$

_____

HRW material copyrighted under notice appearing earlier in this book.

Section	**HOLT PHYSICS**
18-3	**Concept Review**

Capacitance

Use $k_C = 8.99 \times 10^9 \, \text{N} \cdot \text{m}^2/\text{C}^2$.

1. Consider the following units: picofarad, nanofarad, microcoulomb. Explain what quantities they measure, and write their equivalents using powers of 10.

_____

_____

_____

2. A 1.00 pF and a 1.00 nF capacitor each has a charge of 1.00 μC. Which has a higher potential difference between its plates? Show your calculations, and explain your reasoning.

3. A parallel-plate capacitor holds $2.00 \times 10^2 \, \mu$C of charge when a potential difference of 5.00×10^2 V is applied between its plates.

a. What is the capacitor's capacity in units of farads and in units of nanofarads?

_____

_____

b. The potential difference is doubled to 1.000×10^3 V. How does the capacitance change? How does the charge change?

_____

_____

_____

c. How much electrical energy was stored in the capacitor at 5.00×10^2 V? at 1.000×10^3 V?

_____

HOLT PHYSICS
Mixed Review

Electrical Energy and Capacitance

1. A positive charge, q_1, of 5.00×10^{-9} C is placed at $(0, 0)$ in a coordinate system.

   a. Find the potential electrical energy of the two charges when a negative charge, q_2, of 5.00×10^{-9} C is at the following positions in the coordinate system:

   - $(50.0 \text{ cm}, 0 \text{ cm})$ _____

   - $(40.0 \text{ cm}, 30.0 \text{ cm})$ _____

   - $(30.0 \text{ cm}, 40.0 \text{ cm})$ _____

   - $(50.0 \text{ cm}, 0 \text{ cm})$ _____

   - $(-30.0 \text{ cm}, 40.0 \text{ cm})$ _____

   - $(-40.0 \text{ cm}, 30.0 \text{ cm})$ _____

   - $(-50.0 \text{ cm}, 0 \text{ cm})$ _____

   b. Does the electrical potential energy of the two charges increase or decrease when q_2 moves around a circle? Explain.

   _____

   _____

   c. In the space below, sketch the path of the point charge, q_2, in this exercise, and draw the electric force vector acting on it at each of the points indicated in item 1a.

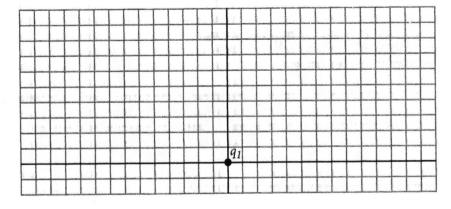

2. Electrons are accelerated in the picture tube of a television through a potential difference of 8.00×10^3 V. (Use the values $q_e = 1.60 \times 10^{-19}$ kg and $m_e = 9.109 \times 10^{-31}$ kg.)

   a. What is the change in the potential energy of each electron traveling in this tube?

   _____

   b. What is the change in the kinetic energy of the electrons?

   _____

   c. At what speed do the electrons hit the screen?

   _____

3. The distance between two vertical plates in a vacuum tube is 6.00 cm. A potential difference of 300 V is applied between the plates. Point A is located 1.00 cm from the positive plate, point B is at 3.00 cm from it, and point C is at 5.00 cm from it.

   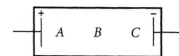

   a. What is the strength of the electric field at points A, B, and C? Is the electric field constant between parallel plates?

   _____

   b. What is the potential difference between the positive plate and points A, B, and C? (Use $\Delta V = Ed$)

   _____

   c. A positive ion with a charge of $+1.60 \times 10^{-19}$ C leaves the positive plate and travels to the negative one. What is its potential energy at the positive plate? at A? at B? at C? at the negative plate?

   _____

4. A 2.00×10^2 nF capacitor has a 4.0×10^1 μC charge.

   a. What is the potential difference between its plates?

   _____

   b. What is the potential energy stored in the capacitor?

   _____

HOLT PHYSICS
Concept Review

Electric Current

1. The sphere of a Van de Graaff generator had 6.00 C of charge. When connected to the ground, it was discharged in 24.0 ms. What was the average discharge current?

_____

2. The current through a light bulb in a flashlight is 0.750 A.

  a. How much charge passed through the filament

   • in 20.0 s? _____

   • in 5.00 min? _____

   • in 2.00 h? _____

  b. How many electrons enter the filament every second?

_____

_____

  c. How many exited the filament every second?

_____

_____

  d. Where do the electrons entering the filament come from? Where do they go after exiting?

_____

3. A battery supplies a 0.015 A current to a small radio. How long should the radio stay on so that 4.80 C passes through each of the following parts of the circuit:

  a. through the battery _____

  b. through the radio _____

  c. through the connecting wires _____

Section
19-2

HOLT PHYSICS
Concept Review

Resistance

1. The label on a small heater specifies its electric performance as 115 V, 4.50 A.

   a. What is the resistance of the heating filament in this heater?

   _____

   b. How much current will it draw when connected to the following:

   • 120 V _____

   • 220 V _____

   • 60.0 V _____

   • 10.0 V _____

2. Three resistors are available for testing a 9.00 V battery. Resistor A has 5.00 kΩ of resistance, resistor B has 5.00 Ω of resistance, and resistor C has 0.0500 Ω of resistance.

   a. How much current will each resistor draw?

   _____

   _____

   b. Which resistor is more useful for testing if the battery is dead? Explain.

   _____

   _____

3. An electrical device of 37.2 Ω resistance performs best when the current is 3.62 A. How much voltage should be applied?

   _____

4. An electronic device performs best with a 1.20 V battery, when the current is between 3.50 mA and 4.20 mA. What is the range of possible resistances for this electronic device?

   _____

Section

19-3

HOLT PHYSICS

Concept Review

Electric Power

1. A food processor draws 8.47 A of current when connected to a potential difference of 110 V.

   a. What is the power consumed by this appliance?

   _____

   b. How much electrical energy is consumed by this food processor monthly (30 days) if it is used on average of 10.0 min every day?

   _____

   c. Assume that the price of electrical energy is 7.00 ¢/kWh. What is the monthly cost of using this food processor?

   _____

2. The electric meter in a house indicates that the refrigerator consumes 70.0 kWh in a week.

   a. What is the power consumption of the refrigerator?

   _____

   b. Assuming it is connected to a potential difference of 120 V, how much current does the refrigerator draw?

   _____

3. The heating element of an electric broiler dissipates 2.8 kW of power when connected to a potential difference of 120 V.

   a. What is the resistance of the element?

   _____

   b. How much current does the broiler draw? Use two ways to find out, and verify your answer.

Chapter	HOLT PHYSICS
19	# Mixed Review

Current and Resistance

1. A 60.0 cm metal wire draws 0.185 A from a 36.0 V battery. Will the current increase or decrease when the following changes are performed? Explain whether the change is due to a change in resistance, a change in potential difference, or other reasons.

   a. The wire is cut into four pieces, and only one segment is used.

   _____

   b. The wire is bent to form an *M* shape.

   _____

   c. The wire is heated to 500°C.

   _____

   d. The 36.0 V battery is replaced by a 24.0 V battery.

   _____

2. A 25 Ω resistance heater is connected to a potential difference of 120 V for 5.00 h.

   a. How much current does the heater draw?

   _____

   b. How much electric charge travels through the heating element during this time?

   _____

   c. What is the power consumption of the heater?

   _____

   d. Use the power and time to calculate how much energy was consumed.

   _____

Chapter

HOLT PHYSICS

19 Mixed Review *continued*

3. The label on a three-way light bulb package specifies 100 W, 150 W, 250 W, 120 V.

   a. How much current does the light bulb draw in each of the three ways? (Assume three significant figures in each of these measurements.)

   _____

   _____

   _____

   b. What is the bulb's resistance in each way?

   _____

   _____

   _____

   c. Compare the cost of using the light bulb for 100.0 h in each way. (Assume that the price is 7.00 ¢/kWh.)

   _____

   _____

   _____

4. An electric hot plate draws 6.00 A of current when its resistance is 24.0 Ω.

   a. What is the voltage across the hot plate's heating element?

   _____

   b. How much power does it consume?

   _____

   c. For what length of time should it be kept on in order to supply 9×10^4 J to a coffeepot? (Assume that all electrical energy is transferred to the coffeepot by heat.)

   _____

Section

20-1

HOLT PHYSICS

Diagram Skills

Schematic Diagrams and Circuits

1. Use the symbols listed in Table 20-1 of the textbook to draw a schematic diagram of an electric circuit that contains one battery, two light bulbs, two resistors, and two switches.

   a. Label the switches *S1* and *S2*. Does either cause a short circuit when closed? Explain.

   _____

   _____

   _____

   b. Add a switch to your diagram, and connect it so that it causes a short circuit when closed. Label it *S3*.

2. A battery, two bulbs, and one switch are placed as shown below. Draw lines representing the wires for connecting these circuit elements so that the following statements will be true.

   a. Both bulbs *A* and *B* are on when the switch is closed.

   b. Only bulb *B* is on when the switch is closed.

   c. Bulb *A* is always on regardless of the switch, and bulb *B* is on only when the switch is closed.

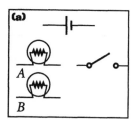 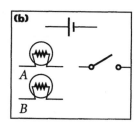 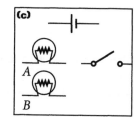

Section
20-2

HOLT PHYSICS

Concept Review

Resistors in Series or Parallel

For each item, sketch a schematic diagram of the circuits and label the components properly.

1. A 12.0 V battery is connected to two resistors in series: $R_1 = 12.00\ \Omega$, $R_2 = 4.00\ \Omega$.

a. Find R_{eq}, the equivalent resistance in this circuit.

_____

b. Find the current through the battery and the current through each resistor.

_____

c. What is the potential difference, ΔV_{eq}, across the equivalent resistance and across each of the resistors?

_____

2. A 12 V battery is connected to two resistors in parallel: $R_1 = 12.00\ \Omega$, $R_2 = 4.00\ \Omega$.

a. Find R_{eq}, the equivalent resistance in this circuit.

_____

b. Find the potential difference, ΔV_{eq}, across the equivalent resistance.

_____

c. What is the current in the equivalent resistance? What is the current in the battery? What is the current in each resistor?

_____

d. What is the potential difference across each of the resistors?

_____

HOLT PHYSICS

20-3 Concept Review

Complex Resistor Combinations

1. The resistors in the circuit below are identical and equal 12.0 Ω. The battery has a potential difference of 24.0 V. Ignore the internal resistance of the battery. (Sketch schematic diagrams of the intermediate circuits as you reduce the complex circuit to a simpler one.)

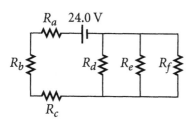

a. Determine the equivalent resistance for this circuit.

_____

b. Find the current in and the voltage across each resistor.

_____   _____

_____   _____

_____   _____

2. Resistor R_f is removed from its present position and connected in series between R_a and the battery.

a. Sketch a diagram of the new circuit.

b. Find the equivalent resistance of the new circuit and the current through each resistor.

_____   _____

_____   _____

_____   _____

Circuits and Circuit Elements

1. Consider the circuit shown below.

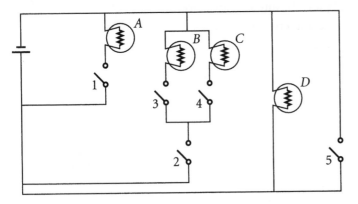

a. Do any of the bulbs have a complete circuit when all the switches are open? Which one(s)? _____

b. Do any of the switches cause a short circuit when closed? Which one(s)? _____

c. Which switches should be kept open, and which should be closed for the following to occur?

   • only bulbs A and B are off _____

   • only bulbs A and C are off _____

   • only bulbs B and C are off _____

2. A light bulb of unknown resistance is connected in series with a 9.0 Ω resistor to a 12.0 V battery. The current in the bulb is 0.80 A.

a. In the space below, sketch a schematic diagram of the circuit.

b. Find the equivalent resistance of the circuit.

_____

c. Find the resistance of the light bulb.

_____

Chapter

20 HOLT PHYSICS
Mixed Review *continued*

3. A light bulb of unknown resistance is connected in parallel to a 48.0 Ω resistor and to a 12.0 V battery. The current through the battery is 2.50 A.

   a. In the space below, sketch a schematic diagram of the circuit.

   b. Find the potential difference across the resistor and across the bulb.

   _____

   c. Find the current in the resistor and in the bulb.

   _____

   d. Find the resistance of the light bulb.

   _____

4. In the circuit below, find the equivalent resistance for the following situations.

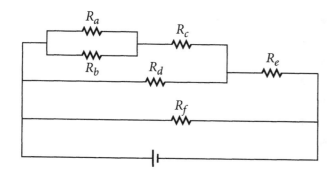

   a. $R_a = R_b = R_c = R_d = R_e = R_f = 10.0 \ \Omega$

   _____

   b. $R_a = 10.0 \ \Omega$; $R_b = 20.0 \ \Omega$; $R_c = 30.0 \ \Omega$; $R_d = 40.0 \ \Omega$; $R_e = 50.0 \ \Omega$; $R_f = 60.0 \ \Omega$

   _____

Section 21-1

HOLT PHYSICS
Concept Review

Magnets and Magnetic Fields

1. You have three marbles, A, B, and C, that look identical. Each of them contains either a magnet or a piece of iron. You have observed that A sticks to B, but B does not stick to C.

   a. Could all three contain iron?

   _____

   _____

   b. Could all three contain magnets?

   _____

   _____

   c. Which of them contain magnets? Which contain iron?

   _____

   _____

2. Many compass needles are placed around a bar magnet at the locations marked on the diagram. Sketch arrows at each point showing to which direction each compass will be pointing.

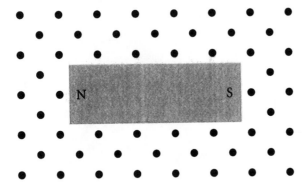

3. In the space below, sketch a horseshoe magnet, and draw lines indicating the direction of the magnetic field around it.

HOLT PHYSICS
Diagram Skills

Electromagnetism and Magnetic Domains

1. Use the convention symbols (×, •, and →) to indicate the direction of
the magnetic field created by electric currents shown in the following
diagrams at points *A, B, C, D, E,* and *F*.

(a)

E $\quad F$

$C \quad | \quad D$

$A \qquad B \quad \uparrow I$

(b)

E

$D \quad \xrightarrow{I} \quad F$

B

$A \qquad\qquad C$

2. How does the strength of the magnetic field at *A* compare with that at *B,
C, D, E,* and *F* in the two situations presented in item 1?

_____

_____

_____

3. The direction of the current is reversed. Sketch the corresponding diagrams, and answer items 1 and 2 again.

NAME ______________________ DATE ____________ CLASS ____________

Section **21-3**

HOLT PHYSICS

Concept Review

Magnetic Force

The charge of an electron is 1.60×10^{-19} C.

1. A proton is moving along the positive *x*-axis with a speed of 1.50×10^5 m/s in a magnetic field of 2.00 T that is oriented along the positive *y*-axis.

a. In the space below, sketch a diagram representing **B** and **v**.

b. Find the direction and magnitude of the electromagnetic force on the proton.

____

____

c. What is the force when the proton moves along the *y*-axis?

____

2. Repeat item 1 for an electron.

3. Repeat item 1 for an alpha particle made of two protons and two electrons.

4. If the magnetic field is uniform along the *y*-axis, do the particles in items 1, 2, and 3 keep moving in a straight line? Describe their path.

____

____

____

Chapter

21 | HOLT PHYSICS
Mixed Review

Magnetism

1. A wire frame carries an electric current in the direction shown. Consider the magnetic field contributed by each segment of the frame at points *A, B, C, D,* and *E.*

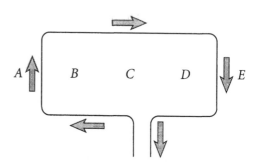

   a. Use the convention symbols (×, •, and →) to represent the direction of magnetic fields created at point *A* by the vertical segments of the frame. Do they have the same direction? the same strength?

   _____

   b. Repeat for the horizontal segments.

   _____

   c. Answer items a and b for points *B, C, D,* and *E,* and fill in the table below.

	leftmost	rightmost	upper	lower
B				
C				
D				
E				

   d. Do the contributions of each segment to the magnetic field cancel out at the center? Explain.

   _____

   _____

   _____

   e. Is the magnetic field resulting from the combined effects of the four sides of the frame stronger inside or outside the frame?

   _____

   _____

2. A 2.0 m long conducting wire has a current of 5.0 in a uniform magnetic field of 0.43 T. The field is parallel to the *x*-axis.

(a) 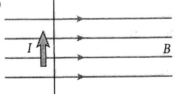 **(b)**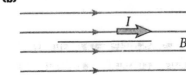

a. What is the force on the wire when it is vertical, parallel to the *y*-axis as shown **a**?

_____

b. What is the force on the wire when it is horizontal, parallel to the *x*-axis as shown in **b**?

_____

3. The wire in item 2 is bent to form a 0.50 m × 0.50 m square carrying the same 5.0 A current, with the positive charges moving clockwise in the frame. The frame is in the same magnetic field (*B* = 0.43 T).

a. Sketch a diagram of the situation. Use arrows to indicate the direction of the current in each segment of the frame.

b. Find the forces acting on each side of the frame. Specify their magnitude and direction.

_____

_____

_____

_____

c. Do the forces on the frame cancel each other? Will the frame be able to move? Will it be able to rotate? Explain.

_____

_____

_____

HOLT PHYSICS

22-1 Concept Review

Induced Current

Consider a loop of wire and a uniform magnetic field as shown below. The loop is shown at five different times as it travels to the right through the magnetic field. The loop is perpendicular to the field.

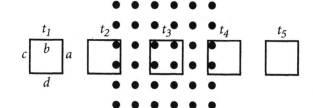

1. Using the right-hand rule for each side (a, b, c, d) of the loop, determine the direction of induced emf for each of the five times above.

   side a: t_1_____ t_2_____ t_3_____ t_4_____ t_5_____

   side b: t_1_____ t_2_____ t_3_____ t_4_____ t_5_____

   side c: t_1_____ t_2_____ t_3_____ t_4_____ t_5_____

   side d: t_1_____ t_2_____ t_3_____ t_4_____ t_5_____

2. Using your answers to item 1, determine the direction (clockwise/counterclockwise) of the current flow for each of the five times.

   t_1 _____ t_2 _____ t_3 _____

   t_4 _____ t_5 _____

3. The loop is a square with sides that are 16.0 cm long, and it is traveling to the right at 8.0 cm/s. The field strength is 1.6 T.

   a. What is the area of the loop?

   _____

   b. How long does it take the loop to completely enter the magnetic field?

   _____

   c. What is the magnitude of the induced emf?

   _____

   d. Find the current in the loop of wire that has a resistance of 0.35 Ω.

   _____

Section 22-2 Concept Review

Alternating Current, Generators, and Motors

Refer to the figure below to answer questions 1–3. Points *A* and *B* represent connections to an external circuit.

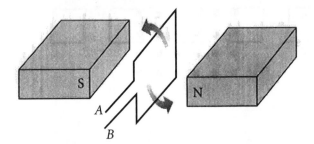

1. In which direction will the loop current flow? (Circle one.)   *A* to *B*   *B* to *A*

2. Suppose you want to *increase* the current. There are several variables to consider. In each case below, choose the appropriate change for each variable. (Circle one.)

    a. Number of loops: Increase   Decrease

    b. Magnetic field strength: Increase   Decrease

    c. Rotational speed: Increase   Decrease

3. The loop shown above is rotating one complete revolution every second. The square loop has sides of 2.5 cm, and the magnetic field strength is 0.75 T. The loop is connected to an 8.0 Ω external circuit.

    a. When (in terms of loop orientation) is induced emf at a maximum?

    _____

    b. When (in terms of loop orientation) is induced emf at a minimum?

    _____

    c. How much time passes (in seconds) between maximum emf and zero emf?

    _____

    d. Using your answers from parts a, b, and c, find the average emf induced in the coil.

    _____

Section

22-3

HOLT PHYSICS

Concept Review

Inductance

Use the figure below to answer the following questions.

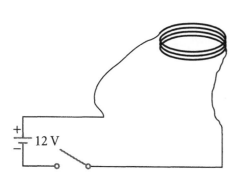
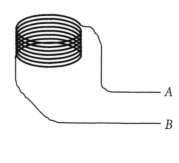

A

B

+
12 V
−

1. Draw the magnetic field created by a clockwise current in the primary loop. Include the area outside of the loop and the part of the field that intersects the secondary loop.

2. Label terminals *A* and *B* of the secondary loop with + or − to indicate the induced emf in the loop when the primary switch is shut. (Hint: consider that the positive terminal will repel the moving positive charge.)

3. If the secondary coil has twice as many turns as the primary coil, calculate the maximum potential difference across the secondary coil—right after the primary coil is "turned on."

_____

4. Explain why the induced emf in the secondary coil is zero when the primary switch has been shut for a long time.

_____

_____

_____

5. When the switch is opened after having been shut for a long time, the primary coil emf goes to zero, but the secondary coil generates a momentary emf. Explain this in terms of changing magnetic fields.

_____

_____

_____

1. Which of the following actions will induce an emf in a conductor?

   a. Move a magnet near the conductor.

   b. Move the conductor near a magnet.

   c. Rotate the conductor in a magnetic field.

   d. Change the magnetic field strength.

   e. all of the above

2. A circular loop (10 turns) with a radius of 29 cm is in a magnetic field that oscillates uniformly between 0.95 T and 0.45 T with a period of 1.00 s.

   a. How much time is required for the field to change from 0.95 T to 0.45 T?

   _____

   b. What is the cross-sectional area of one turn of the loop?

   _____

   c. Assuming that the loop is perpendicular to the magnetic field, what is the induced emf in the loop?

   _____

3. Electric generators convert mechanical energy into electrical energy.

   a. What are the requirements for generating emf?

   _____

   _____

   _____

   b. The mechanical energy input is usually rotational motion. What are two possible sources of rotational motion?

   _____

   _____

4. A 250-turn generator with circular loops of radius 15 cm rotates at 60.0 rpm in a magnetic field with a strength of 1.00 T.

   a. What is the angular speed of the loops?

   _____

   b. What is the area of one loop?

   _____

   c. What is the maximum emf?

   _____

   d. What is the rms emf?

   _____

5. An electric motor is sometimes called a generator in reverse. Explain your understanding of this statement.

   _____

   _____

6. Consider a two-coil transformer joined by a common iron core.

   a. If the current in the primary side is increased, what happens to the magnetic field in the core?

   _____

   _____

   b. What effect does the answer to item 6a have on the secondary coil?

   _____

   _____

   c. Fully explain the effect of reducing the current to the primary side of a transformer.

   _____

   _____

Section
23-1

HOLT PHYSICS
Concept Review
Quantization of Energy

1. According to the classical theory of physics, the energy radiated by a
   blackbody approaches infinity as the wavelength of the emitted light
   approaches zero.

   a. Why was this considered a problem for classical physics?

   _____

   _____

   b. Max Planck solved this problem in 1900. What was the key to the
   solution?

   _____

   _____

   c. How does Planck's assumption solve the "ultraviolet catastrophe"?

   _____

   _____

   _____

2. A ringing bell oscillates at 440 Hz.

   a. How much energy (in joules) is carried away in a one-quantum
   change of this system?

   _____

   b. Convert your answer to units of electron-volts.

   _____

3. The equation for the maximum kinetic energy of an ejected photoelectron is $KE_{max} = hf - hf_t$.

   a. Rearrange this equation to solve for the work function.

   _____

   b. If photoelectrons with 2.55 eV of maximum kinetic energy are observed
   when a 1.17×10^{15} Hz light is used, find the work function of the metal.

   _____

Section
23-2

HOLT PHYSICS
Concept Review

Models of the Atom

1. Write a brief description of Rutherford's model of the atom.

   _____

   _____

   _____

   _____

2. Why was Rutherford surprised that some of the alpha particles were scattered backwards?

   _____

   _____

3. Even though some atoms were scattered backwards, why did Rutherford conclude that the atom was mostly empty space?

   _____

   _____

4. A major problem with Rutherford's model is that atoms would quickly collapse rather than continue to exist (as we know from observation of the everyday world). Explain in terms of energy why the Rutherford atom would collapse.

   _____

   _____

   _____

   _____

Section
23-3

HOLT PHYSICS
Concept Review
Quantum Mechanics

1. Light acts as both a wave and a particle.

  a. Give an example in which light acts like a wave.

  _____

  _____

  b. Give an example in which light acts like a particle.

  _____

  _____

2. Heisenberg's uncertainty principle states that it is impossible to simultaneously measure both the position and the momentum of an object with complete certainty. Explain why this uncertainty is a big concern when conducting measurements on a small object, such as an electron, but is not a consideration when measuring the position and momentum of a large object, such as an athlete. (Hint: Consider the amount of uncertainty relative to the size of the measured value.)

  _____

  _____

3. Calculate the de Broglie wavelength for the following objects:

  a. a 1550 kg car moving at 29.1 m/s _____

  b. a 90 800 kg ship moving at 13.5 m/s _____

  c. a 75 kg person moving at 10.5 m/s _____

  d. an 8.2 kg baby crawling at 2.2 m/s _____

4. In terms of the uncertainty principle, how was the quantum mechanical model of the atom an improvement over Bohr's model?

  _____

  _____

  _____

  _____

Chapter	**HOLT PHYSICS**
23	**Mixed Review**

Atomic Physics

1. The photoelectric effect does not occur below the threshold frequency, which corresponds to the work function of the metal. Using the concept of quantization of light, explain why this is true.

_____

_____

2. Why is the maximum kinetic energy of a photoelectron always less than the energy of the photon that ejected the electron?

_____

3. **a.** Some of the alpha particles in Rutherford's experiment were scattered backwards. What conclusion was drawn from this observation?

_____

   b. Most of the alpha particles continued through the foil almost completely undisturbed. What is implied by this observation?

_____

4. De Broglie proposed that all matter has wavelike properties, and electrons have been observed to diffract and exhibit other wavelike properties when passed through a slit.

   a. Calculate the de Broglie wavelength of an electron moving at 5.0×10^4 m/s.

_____

   b. Calculate the de Broglie wavelength of a 25 g ball moving at 5.0×10^1 m/s.

_____

   c. Explain why you do not observe wavelike properties for objects such as the ball in part b.

_____

_____

5. a. State the uncertainty principle.

_____

_____

b. Explain why the uncertainty principle supports the theory of an electron cloud rather than a distinct orbit for electrons.

_____

_____

6. The accuracy of measuring an electron's position and momentum around a nucleus is limited by the change caused by the measuring instrument—the reflection of light photons. The measurement of a planet's position and momentum around the sun is not limited. Explain the difference in terms of the effect of the light used to create an image of the electron and the planet.

_____

_____

_____

7. What is the threshold frequency of a metal whose work function is 4.82 eV?

_____

8. Describe the effect of shining a light that has a frequency below the threshold frequency for a given surface.

_____

_____

9. If the energy deposited by light does not eject electrons, where does it go? (Hint: Consider other parts of an atom.)

_____

_____

10. How would the energy accumulation in item 9 be observed?

_____

_____

Section
24-1

HOLT PHYSICS
Concept Review

Conduction in the Solid State

1. Beside each of the following properties, identify the type of material associated with the property. Circle all that apply.

    a. low resistance to electron flow   insulator   conductor   semiconductor

    b. high resistance to electron flow   insulator   conductor   semiconductor

    c. conduction and valence bands overlap   insulator   conductor   semiconductor

    d. large energy gap between bands   insulator   conductor   semiconductor

    e. small energy gap between bands   insulator   conductor   semiconductor

2. In terms of the size of the energy gap between the valence and conduction bands, explain why it is easier to cause a semiconductor to conduct electricity than an insulator.

_____

_____

_____

3. For a material to conduct electricity, there must be electrons in the conduction band. Conducting materials have electrons in the conduction band, while semiconductors and insulators normally do not. However, semiconductors and insulators can have electrons in the conduction band if the electrons undergo transitions to higher levels. Discuss different ways of exciting electrons into the conduction band for insulators and semiconductors.

_____

_____

_____

4. An isolated atom does not have energy bands; it has energy levels. Why do we consider energy bands when discussing properties of materials?

_____

_____

_____

Section
24-2

HOLT PHYSICS
Concept Review
Semiconductor Applications

1. When an electron moves into the conduction band in a semiconductor, it leaves behind a hole in the valence band.

   a. Is it easier for a neighboring electron to move to the hole in the valence band or to the conduction band?

   _____

   b. Explain the importance of this hole in terms of the conduction of electricity in the semiconductor.

   _____

   _____

   _____

2. Silicon is a commonly used semiconductor. It has four valence electrons.

   a. In order to make a p-type semiconductor, how many valence electrons should the doping material have?

   _____

   b. Does this doping material cause the semiconductor to become positively charged? Why or why not?

   _____

   _____

   c. How many valence electrons should an n-type doping material have?

   _____

   d. Does this cause the semiconductor to become negatively charged?

   _____

Section
24-3

HOLT PHYSICS
Concept Review

Superconductors

1. A primary cause of resistance in materials is the thermal vibration of the atoms in the lattice structure. However, even at absolute zero, many materials still have some resistance to electric current. What is the cause of this residual resistance?

   _____

   _____

2. In the BCS theory of superconductivity, electrons travel in pairs through a lattice.

   a. What happens to the positively charged lattice atoms as one electron passes near those positive charges?

   _____

   _____

   _____

   b. What effect does the change in the lattice have on the second electron in the pair?

   _____

   _____

   _____

   c. The first electron loses some momentum while interacting with the lattice. Where does this momentum end up?

   _____

   d. Imagine that we could positively identify a Cooper pair. If we were to watch them travel through the lattice, would we see the pair travel together through the entire lattice? Explain your answer.

   _____

   _____

   _____

Mixed Review

Modern Electronics

1. In the space below, draw diagrams of the valence and conduction bands for an insulator, a semiconductor, and a conductor. Include the relative size of the energy gap.

2. Why do conductors and semiconductors allow current to flow more easily than insulators do?

_____

_____

3. Individual atoms have energy levels, not bands. What causes energy bands to form in a solid?

_____

_____

4. What are two methods for exciting electrons into the conduction band in semiconductors?

_____

_____

5. Two electrons ordinarily repel each other. How is it possible to have electrons bound together in a Cooper pair?

_____

_____

6. How is the construction of a transistor different from the construction of a diode?

_____

_____

Chapter **24**	**HOLT PHYSICS** **Mixed Review** *continued*

7. a. When doping a semiconductor, what property is important?

_____

b. How does doping a semiconductor with an impurity increase the
semiconductor's conductivity?

_____

_____

_____

8. Explain the difference between p-type and n-type semiconductors in
terms of charge carriers and doping.

_____

_____

_____

9. Why does a diode allow current in one direction and resist current in the
other direction?

_____

_____

_____

10. How are superconductors different from conductors and semiconductors?

_____

_____

_____

11. A superconducting ring can be used as a storage device, while a conducting ring cannot. Explain the difference. Where does the energy go in a
nonsuperconducting ring?

_____

_____

_____

The Nucleus

1. A certain atom has eight protons, eight electrons, and eight neutrons.

   a. How many nucleons does this atom have?

   _____

   b. What is the atomic number of this atom?

   _____

   c. What is the mass number of this atom?

   _____

   d. If the nucleus of this atom has a mass of 16.124 552 u, calculate the binding energy of the nucleus.

   _____

   e. What is the significance of the binding energy?

   _____

   _____

   f. Would an atom with eight protons, eight electrons, and nine neutrons be a different element? Explain.

   _____

   _____

2. Two protons in a nucleus experience a very large repulsion force.

   a. What prevents these two protons from accelerating away from each other?

   _____

   b. As a nucleus gets larger, what happens to the ratio of protons to neutrons?

   _____

   _____

Section
25-2
HOLT PHYSICS
Concept Review

Nuclear Decay

1. List and describe the three types of radiation emitted by radioactive materials.

   _____

   _____

   _____

2. Find the element produced in the following decays:

   a. Nitrogen-17 decays by emitting a beta particle. _____

   b. Uranium-235 decays by emitting an alpha particle. _____

   c. Uranium-238 decays by emitting a beta particle. _____

   d. Plutonium-239 decays by emitting an alpha particle. _____

3. What does the term *half-life* mean?

   _____

   _____

4. What is the decay constant?

   _____

   _____

5. What is the mathematical relationship between the decay constant and the half-life of a substance?

   _____

   _____

6. Find the decay constant of a material that has a half-life of 14 s.

   _____

7. Find the half-life of a material that has a decay constant of 2.20×10^{-8} s^{-1}.

   _____

8. How much of the material in item 7 will remain after two years?

   _____

Concept Review

Nuclear Reactions

1. A typical nuclear reaction is $_{0}^{1}n + _{92}^{235}U \rightarrow _{56}^{141}Ba + _{36}^{92}Kr + 3\,_{0}^{1}n$.

   a. Is this a fission reaction or a fusion reaction?

   _____

   b. What are the reactants in this reaction?

   _____

   c. What are the products of this reaction?

   _____

   d. Are mass and charge conserved in this reaction?

   _____

   e. This reaction produces three neutrons. What might happen if each
   neutron is absorbed by another uranium nucleus?

   _____

   _____

   f. What is the danger of an uncontrolled nuclear reaction?

   _____

   _____

2. Another possible reaction is $_{1}^{1}H + _{2}^{3}He \rightarrow _{2}^{4}He + _{1}^{0}e + \nu$.

   a. Is this a fission reaction or a fusion reaction?

   _____

   b. What are the reactants in this reaction?

   _____

   c. What are the products of this reaction?

   _____

   d. Are mass and charge conserved in this reaction?

   _____

Section
25-4

HOLT PHYSICS
Concept Review

Particle Physics

1. List the four fundamental interactions in order of relative strength.
   Describe each interaction, including relative strength, effects, and the
   range of force.

   _____

   _____

   _____

   _____

2. The four fundamental interactions each have a mediating particle.

   a. List the mediating particles for each of the following types of
      interactions:

      gravitational _____

      weak _____

      electromagnetic _____

      strong _____

   b. Which mediating particle has not yet been discovered?

   _____

3. The standard model proposes the existence of a particle called the
   Higgs boson.

   a. What is the reason scientists predict the existence of the Higgs boson?

   _____

   _____

   _____

   b. Why has this particle not been observed?

   _____

   _____

   _____

Subatomic Physics

1. Determine the number of neutrons in the following nuclei:

a. $^{235}_{92}U$ _____

b. $^{238}_{92}U$ _____

c. $^{239}_{93}Pu$ _____

d. $^{2}_{1}H$ _____

e. $^{3}_{1}H$ _____

f. $^{14}_{6}C$ _____

g. $^{17}_{7}N$ _____

h. $^{40}_{18}Ar$ _____

2. Consider the following pairs of nuclei: $^{12}_{6}C$, $^{13}_{6}C$ and $^{238}_{92}U$, $^{239}_{93}Pu$.

a. What does the first pair have in common?

_____

b. What is the difference between the nuclei in the first pair?

_____

c. What does the second pair have in common?

_____

d. What is the difference between the nuclei in the second pair?

_____

e. Describe the similarities between the two pairs.

_____

_____

f. Describe the differences between the two pairs.

_____

_____

Chapter	HOLT PHYSICS
25	**Mixed Review** *continued*

3. A nucleus decays by emitting a beta particle.

    a. Compare the atomic mass of the new nucleus with that of the original nucleus.

    _____

    _____

    b. Compare the atomic number of the new nucleus with that of the original nucleus.

    _____

    _____

    c. Which nucleus would you expect to have a larger binding energy? Explain.

    _____

    _____

    d. Which nucleus would have a larger mass defect? Explain.

    _____

    _____

4. Fusion in the sun creates high temperatures that tend to make the sun expand. What keeps the reaction contained?

    _____

    _____

5. A deuteron, ^2_1H, may decay. Could it decay by emitting an alpha particle? Explain.

    _____

    _____

6. What two quantities must be conserved in a nuclear reaction equation?

    _____

    _____